# A Textbook of Vermicompost: Vermiwash and Biopesticides

**The Authors**

**Dr. Keshav Singh**, Lecturer, Department of Zoology, D.D.U. Gorakhpur University, Gorakhpur, (U.P.), INDIA, born on 26th January, 1970. M.Sc. Zoology with specialization in Entomology in 1991, Ph.D. in Zoology in 1997 title of the thesis "Studies on molluscicidal activity of *Azadirachta indica* A. Juss (Neem) against harmful gastropods". More than 35 research papers published in journals of international repute and written two books (Enzyme inhibition by different bait formulation against the snail *Lymnaea acuminata*. Environmental pollution and toxicology Eds. (B.D. Joshi, P.C. Joshi and Namita Joshi) A. P. H. Publishing Corporation, India, 2008 and Vermicomposting: A Boon for Soil, Plant and Environment, Lambert publishing Academy, Germany, 2011). Worked as Senior Research Fellow in project of ICAR, New Delhi (1995-1996) and DBT, New Delhi (1996-1999). Presented 5 research papers in National seminars/ conferences/symposia. Teaching experiences 18 years in UG and PG classes. Three students have been awarded Ph. D. degree and four students are currently registered for Ph. D. degree under my supervision. One major research project (UGC New Delhi) completed. Actively engaged in research activities for the last twenty years in field of Toxicology, Biochemistry and Phermacology. Worked for Ph.D. degree on control of harmful gastropods by using the plant product biopesticides. The harmful gastropods; *Lymnaea acuminata* and *Indoplanorbis exustus* are the intermediate hosts of *Fasciola hepatica* and *F. gigantica* causes faciolisasis in cattle population. The use of natural molluscicides for control of gastropods are safe for the environment, less expensive, easily available and easily biodegradable. For last eight years actively working on vermibiotechnology, waste management by using earthworm *Eisenia foetida*. Vermicomposts and vermiwash with biopesticides have a significant effect on the growth, productivity and pest infestation. Organized two formers awareness programme on preparation and use of vermicompost and vermiwash and their benefit. I have been Assistant Proctor, Assistant Dean Student welfare, Chairperson of Athletics Association, admission committee and at present, Programme Officer in National Service Scheme (NSS), Chair person of athletics Association and Assistant Proctor of Deen Dayal Upadhyaya, Gorakhpur University, Gorakhpur.

**Dr. Gorakh Nath** was born on 20 October, 1981. He passed M.Sc. in Zoology with specialization in Entomology in 2006. Ph.D. in Zoology, in year 2009 in the field of Vermibiotechnology, Department of Zoology, Deen Dayal Upadhyay Gorakhpur University, Gorakhpur (U.P.), India. He is working as Lecturer in S. V. M. Mahavidhyalay Aryanagar Gorakhpur and two year teaching experience in UG and PG level. He has been published 11 research papers in International repute Journal, 14 abstracts; and attended eight National/ International Conferences/Seminars/Symposium as well as published one book entitled "Vermicomposting: A Boon for Soil, Plant and Environment" Lambert publishing Academy, Germany in year 2011.

Dr. Rabish Chandra Shukla was born on 01 July, 1981. He working in Vermibiotechnology field, Department of Zoology, Deen Dayal Upadhyay Gorakhpur University, Gorakhpur (U.P.), India and awarded Ph.D. degree in 2010, passed M.Sc. in Zoology with specialization in Entomology in 2005. He appointed as Lecturer in S. P. Institute of Sci. and Tech and three year teaching experience and published his four research papers in International repute Journal as well as attended two National Seminars and symposium.

**Dr. D. K. Bhartiya** working on vermibiotechnology, and awarded Ph.D degree in 2013 from Department of Zoology, Deen Dayal Upadhyay Gorakhpur University, Gorakhpur (U.P.), India. M.Sc. Zoology with specialization in Entomology in 2008. He awarded Senior Research Fellow (SRF) of Rajive Gandhi National Fellowship, UGC, New Delhi. Published four research papers in Journal of International repute and one Book entitled "Vermicomposting: A Boon for Soil, Plant and Environment" Lambert publishing Academy, Germany in year 2011, Attended twelve National Seminars/Workshop.

# A Textbook of Vermicompost: Vermiwash and Biopesticides

**Dr. Keshav Singh**
*Lecturer, Department of Zoology,*
*D.D.U. Gorakhpur University, Gorakhpur – 273 009 (U.P.), INDIA*

**Dr. Gorakh Nath**
*Department of Zoology,*
*S.V.M.M.V. Aryanagar, Gorakhpur, (U.P.), INDIA*

**Dr. Rabish Chandra Shukla**
*Department of Zoology,*
*Sardar Patel Science and Technology Mahavidyalay,*
*Moharipur, Gorakhpur, (U.P.), INDIA*

**Dr. Deepak Kumar Bhartiya**
*Ph. D. Scholar*
*Department of Zoology,*
*D.D.U. Gorakhpur University, Gorakhpur – 273 009 (U.P.), INDIA*

**2025**

**BIOTECH BOOKS®**

ISBN 978-81-7622-322-5



*Published by*: **BIOTECH BOOKS®**
4762-63/23, Ansari Road, Darya Ganj,
New Delhi - 110 002
Phone: 09899295259, 011-43003222
E-mail: biotechbooks@yahoo.co.in
Website: www.biotechbooks.in

*Typeset at*: **Classic Computer Services**
Delhi - 110 035

*Printed at*: **Replika Press Pvt. Ltd**

PRINTED IN INDIA

# Preface

Continuous applications of chemical fertilizers disturb the soil texture, physico-chemical properties as well as affect the human health and environment. Nutritional component like carbohydrate, amino acid and ascorbic acid are reduced in the food stuffs by use of nitrogenous, phosphatic fertilizer in soil. Biological wastes are a serious problem for society. It caused environmental hazards and various ill effects on the human life and their domesticated animals if their proper management and disposal practices are not available. Use of organic matter in the form of vermicompost or vermiwash in agriculture field increases the bioavailability of nutrient to the growing plant.

The vermicomposting could be an adequate technology for management of biological wastes through earthworms. *Eisenia fetida* is one of the suitable species for vermicomposting because it can tolerate wide variation of ecological factor such as temperature, humidity and wide variety of wastes.

It is believed that the knowledge from this book will give not only information of various useful effect of vermicompost and vermiwash for agricultural crops but also help in promotion of organic fertilizer by management of biological wastes and minimized the environmental pollution caused by the use of chemical fertilizers.

*Dr. Keshav Singh*

*Dr. Gorakh Nath*

*Dr. Rabish Chandra Shukla*

*Dr. Deepak Kumar Bhartiya*

# Contents

# 1

# Introduction

Increased demand of agricultural products boosted tremendous pressure for more production of grains by adopting modern agricultural practices. Modern agricultural practices contribute much deleterious effect on the field soil (Manning, 2000; Peng *et al.,* 2006). Organic matters are not fully recycled, which affect the nutrient regenerative capacity of soil and ultimately to crop production (Soytong and Soytong, 1996; Setboonsarng and Gilman, 1999; Soytong *et al.,* 1999; Adriano, 2001; Wezel and Rath, 2002). Use of high chemical fertilizers for more production of crops is one of the major caused of destruction of soil flora fauna, responsible for their natural quality (Lasat, 2002; Peyvast *et al.,* 2008).

Higher use of chemical fertilizers leads to high concentration of some chemicals and metals, which ultimately affect the crops and watershed (Eghball and Gilley, 1999; Kpomblekou *et al.,* 2002). Such agricultural practices are dangerous for soil fertility and conservation that may lead to desertification of land (Brady and Weil, 2002). Nutrient like protein, amino acid, vitamins and ascorbic acid etc. are reduced due to excessive use of phosphatic nitrogenous and potash fertillizers in soil. Besides this, the nitrogenous fertilizers pollute the water and food items, causing serious health problems (Bhattacharya, 2004). High nitrate ($NO_3^-$) concentration in water and foodstuff can cause of gastric cancer in human body (Trivedi and Goel,1984). These chemical fertilizers show a spectacular response but for only short period as well as for single crop. In forthcoming the need of fertilizers quantity to be increased for get high yield target. The productivity of crops and nature of the soil is affected with the uncountable utilization of synthetic fertilizers (Gupta, 2005). Irrigation without proper arrangement of drainage water the soil may be gets alkaline or saline (Fulekar and Jadia, 2008).

Now there is a growing realization that adoption of ecological and sustainable farming practices can only reverse the decline trend in the global productivity and environmental protection (Bulluck and Ristaino, 2002). In the developing countries, where the high cost limits the use of chemical fertilizers. Indian agriculture practices have tremendous value for sustainable agriculture (Singh, 2002; Yang *et al.,* 2005).

The agricultural scientists have focused the attention on the development of conventional system of agriculture which is chemical free and safe for human being as well as animals (Gupta and Garg, 2007). It embraces several forms of non-conventional agriculture practices called as organic farming. Organic farming through vermicomposting is the pathway that leads us to live in harmony with nature. Organic farming is the key to a minimized environmental pollution, conserve soil fertility, and check the soil erosion and use of non-renewable natural resources through implementation of appropriate conservation principles (Reganold *et al.,* 2001; Bisoyi, 2003; Gupta, 2005).

The organic farming systems are capable to maintain the soil fertility with improving quality of agricultural products, conserve resources and socially supportive. Organic farming is a biotechnological process, which could provide a solution to tackle the immediate need of plant nutrient for sustainable productivity. The uses of biofertilizers are supposed to be environmentally safe and best for sustainable agriculture (Tiwari *et al.,* 1989; Wani *et al.,* 1995). Usually most of the biofertilizers for example, vermicomposts are prepared from biological wastes. Before discuss the vermicomposting; it is better to know about biological wastes.

# 2

# Biological Wastes

The biological wastes caused environmental hazards and various ill effects on human life and their domesticated animals, if their proper management and disposal practices are not available (Bhattacharya and Chattopadhyay, 2004). High rate of industrializations has also increased the problems of solid waste management. These problems starts from rural level *i.e.* agro, kitchen, vegetable wastes and animal wastes etc. as well as to industrial and urban level *i.e.* solid wastes of textile mill, sugar mill, vine industries, dairy plant sludge and municipal solid wastes etc. These wastes are harmful to human being and their cattle, causing several diseases, odour and pollution problems through microbial decomposition (Reinecke *et al.*, 1992; Butt, 1993).

Gaur gum (*Cyamposis tetragonaloba*) industrial wastes caused liberation of harmful gases which caused various odour and serious environmental problems (Reinecke *et al.*, 1992; Mitchell, 1997; Gunadi *et al.*, 2002; Gunadi and Edwards, 2003; Suthar, 2006a, b; Suthar, 2007b). The production of fly ash on large scale is a serious problem for human health and environment (Bhattacharya and Chattopadhyay, 2004; Venkatesh and Eevera, 2008). The fly ash is fine residue flow exhausts when coal is burn in power station and creating acute waste disposal problems (Kalra *et al.*, 1997; Stevinus and Dunn, 2004; Venkatesh and Eevera, 2008). The problem of post harvest residue of some local crop *e.g.* wheat (*Triticum aestivum*), pearl millet (*Pennisetum typhoides*), sorghum (*Sorghum vulgare*), golden gram (*Vigna rediata*), rice stubble (Talashilkar *et al.*, 1999) and water hyacinth (*Eichhornia crassipes*) (Gajalakshmi *et al.*, 2001) are also the major waste problem in the field (Kaushik *et al.*, 2003; Muthukumaravel *et al.*, 2008; Suthar, 2008 a). Bansal and Kapoor (2000) has reported the management of various agricultural residue and crop waste through the use of earthworms. The municipal solid wastes, city garbage and street sweeping wastes cause odour problems (Fredrickson *et al.*, 1997; Garg *et al.*, 2006a). The livestock excreta (cow, buffalo, horse, goat and sheep dung etc.) create noxious problems if they are not managed properly (Nogales *et al.*, 1999; Garg *et al.*, 2005; Loh *et al.*, 2005). The average availability of these waste are 350 mt roughly (Verma and Sharma, 2000). The microbial decompositions of these wastes produced various harmful gases (Garg *et al.*, 2006a).

The various kind of industrial wastes like- vine fruit industries sludge (Atharasopoulous, 1993), dairy processing plant sludge (Gratelly *et al.*, 1996; Elvira *et al.*, 1998), paper mill sludge (Elvira *et al.*, 1997; Gajalakshmi *et al.*, 2002), sewage sludge (Fang *et al.*, 1999) and sugar plant litter (Fredrickson *et al.*, 1997) cause lot of problems on the agricultural land. Different natural and anthropogenic waste has been converted in the useful compost by different species of earthworm (Benitez *et al.*, 1999; Ghosh *et al.*, 1999; Reddy and Reddy, 2000; Eastman *et al.*, 2001; Maboeta and Van-Rensburg, 2003).

Organic farming through vermicomposting is a better option for management of wastes by the earthworms and improvement of soil quality. It is one of the interesting aspects, since it contribute to a broad relationship among food, environmental quality and safety of human and animal health (Parr *et al.*, 2002). It converts the waste in to a valuable product as "vermicompost", which is used in agriculture as manure to increase the production of food. Recycling of wastes through vermicomposting reduced the problem of wastes. Vermicomposting is found to be a suitable way for the proper management and potential use of wastes.

# 3

# Vermicomposting

Vermicomposting is an adequate technology for bio-oxidation and stabilization of organic materials with joint action of earthworms and microorganisms where, organic wastes are converted into nutrient rich plant growth media. However, the involvement of earthworms in vermicomposting decreases the time of stabilization of the waste and produce rich organic nutrient as vermicompost. Although, microbes are responsible for biochemical degradation of organic matter while, earthworm are important for the vermicomposting, conditioning of the substrate and alter the biological activity. The vermicomposting is an aspect involving the use of earthworm as natural bioreactors for effective recycling of non-toxic organic wastes for soil. They effectively increase the beneficial soil micro-flora, destroy the pathogen of soil and convert organic wastes in to valuable products (Benitez *et al.*, 2005; Garg and Kaushik, 2005; Suthar, 2006 a, 2007b).

Earthworms play an important role in stabilization of inorganic plant nutrients to organic form and increased the soil fertility. The worms added their cast with compost and increased the inorganic nutrients many times along with some plant growth hormones and vitamins. They also help in the problems of management of rotting solid organic waste in dumping site which caused odour problems, polluted the soil, water, air and health of human population. Extensive scientific researches have provided sufficient knowledge regarding the usefulness of earthworms (Buche, 1977; Levi-Minzi *et al.,* 1986). According to Charles Darwin that "the worms seem to be the great promoters of vegetation, perforating and loosing the soil, rendering the pervious to rains and the fiber of plant by drawing straw and stalk of leaves and twinges in to it. Most of all by throwing up infinite number of lumps of earth called worm-cast, which being their excrements is a fine manure for grain and grass" (Gupta, 2005).

Vermicomposting could be an adequate technology for management of organic wastes in to valuable product through earthworm (Vermi Co., 2001; Crescent, 2003; Elvira *et al.,* 1997). During the process the important plant nutrients such as nitrogen, potassium, phosphorus, calcium etc. present in feed materials are converted in to more soluble and absorbable form for plant microbial activity than those in parent substrate (Ndegwa and Thompson, 2001). Vermicompost is a peat like material egested by earthworms after ingestion of various solid organic wastes; it is much fragmented, porous and suitable for microbial activity than original waste due to high degree of humification and decomposition (Wani, 2002; Edwards and Bohlen, 1996). Aalok *et al.* (2008) have reported that vermicomposting is a better option for management of solid organic wastes. Vermicompost contain biologically active substances such as plant growth regulators through vermistablization of industrial sludge by worms (Tomati *et al.,* 1987).

Vermicomposting is an easy and effective way to recycle agriculture wastes, city garbage and kitchen wastes along with bioconversion of organic waste materials in to nutritious compost by earthworm activity. Not only these plant nutrient growth regulator but also have beneficial bacterial and actinomycetes population in large number of worms eggs, which hatch out within a month is equivalent to a minifertilizer factory in the soil (Mall *et al.,* 2005). It increases the porosity aeration, drainage water holding capacity, which reduced the irrigation water requirement for crops. It improves nutrients availability and could act as complex fertilizer granules (Purohit, 2003; Yadav, 2003).

Vermicomposting involves great reduction in the population of harmful pathogenic microorganism. It is accepted that the thermophilic stage during the composting process eliminates the pathogenic organisms. In aerobic process or composting through microbes leads to the mineralization. However, composting through earthworm increases and accelerates the nitrogen mineralization rate (Bansal and Kapoor, 2000; Kaushik and Garg, 2003; Atiyeh *et al.*, 2000). The humification process that takes place during the maturation stage of vermicomposting is greater and faster. This process may also bring about greater decrease of bio-available heavy metal then in composting process. Vermicomposting is more applicable than composting in term of waste management. It is one of the most important procedure in order to stabilized organic wastes at the same time managing to the environmental problem (Gupta, 2005).

Vermicomposting are finely divided peat like materials with high porosity aeration, drainage and water holding capacity (Edwards and Burrows, 1988). Vermicomposts has been shown to have higher levels of organic matter organic carbon total and available NPK and other micronutrient, microbial and enzyme activity and plant growth regulator (Mulongoy and Bedoret, 1989; Tomati and Galli, 1995; Edwards and Bohlen, 1996; Parthasarathi and Ranganathan, 2002; Chaoui *et al.*, 2002). Vermicomposts have less soluble salts, neutral pH, greater cataion exchange capacity and humic acid content (Albanell *et al.*, 1988). The content nutrients are taken up by the plants readily. Vermicomposts are more stable than their parent material, with higher availability of nutrient and improved physico-chemical properties of soil (Azarmi *et al.*, 2008).

Vermicomposts can be produce using animal manure as substrate (Contreras-Ramos *et al.*, 2004). Vermicomposts promote soil properties such as granulation, fine filth, efficient aeration, easy root penetration and improved water holding capacity (Edwards and Bohlen, 1996; Peyvast *et al.*, 2008). Vermicomposts constantly promote the biological activity which can cause plants to germinate, flower, growth and yield of bedding plant container media in dependent of nutrient availability (Atiyeh *et al.*, 2000; Arancon *et al.*, 2004). Organic manure can increased soil organic matter, soil water retention and transmission and other physical properties of soil, decreased bulk density, and increased aggregation (Turner *et al.*, 1994; Zebrath *et al.*, 1999).

Vasanthi and Kumaraswamy (1999) studied that application of vermicomposts that not only enhanced soil fertility but also improved the yield of crop. Long term manure application is often recommended to maintain soil fertility (N'Dayeagmiye, 1990). Vermicomposts had been shown to influence the growth and productivity of plant, cereals, legumes, sugarcane (Rajkhowa *et al.*, 2000; Lalitha *et al.*, 2000), vegetable (Atiyeh *et al.*, 1999; 2000); ornamental and flowering plant (Atiyeh *et al.*, 2000), field crops (Mba, 1996; Buckerfield and Webster, 1998). Atiyeh *et al.* (2001) and Chaoui *et al.* (2003) have shown that vermicompost when used in bedding media improve productivity. They have further shown that the grater response from the plant could be observed only when the vermicompost was used at 10-40 per cent of the amount in growth medium, contrarily; higher doses of vermicompost do not increase plant growth, because of high salt content. Garg and Bhardwaj (2000) have studied the effect of vermicomposts of parthaneum on the growth and yield of wheat.

Humankind has the knowledge from ancient time about the ability of earthworm to transform the organic wastes in to value added product with soil fertilizing capabilities. In 1881, Charles Darwin's regarded as the first modern scientific study of the beneficial role of earthworm in soil ecology. The United State of America in the early part of the 20th century was regarded for the beginning of vermicomposting. By the late 1940s, earthworm growers were promoted as an effective method for farmers to improve their soil fertility on the basis of scientific use of vermicomposting (Bouche, 1987; Kale, 1991). In the 1970s, researchers started the exploring of earthworm potential to reduced waste disposal problems and soil reclamation in the USA, Japan, Britain and France. The home-scale vermicomposting was rapidly increased in the popular industrialized country after 1980s. Since 1989, Cuba was heavily depends on the vermicomposting for disposal of agricultural and municipal solid wastes and as primary source of soil fertility. Thus, the large scale experimentation determined the vermicomposting is a high potential of waste management technology (Wong and Griffith, 1991; Bhole, 1992; Atiyeh *et al.*, 2000; Eastman *et al.*, 2001).

Vermicomposting could be an adequate technology for the transformation of wastes into valuable products (Elvira *et al.*, 1997;

Appelhalf *et al.*, 1998; Nagavallemma *et al.*, 2004). During the vermicomposting process the important nutrients like N, P, K, and Ca present in the feed material are converted into much soluble nutrients for plant through earthworm action (Ndegwa and Thompson, 2001). Vermicomposting has been reported to be a viable, cost-effective and rapid technique for the management of the domestic animals as well as industrial wastes in to value added material (Payal *et al.*, 2006). Earthworms play an important role in stabilization of inorganic plant nutrients to organic form and increased the soil fertility (Ranganathan, 2006). The worms added their cast with compost and increased the inorganic nutrients many times along with some plant growth hormones and vitamins (Atiyeh *et al.*, 2002). They also help in the problems of management of rotting solid organic waste in dumping site which caused odour problems, polluted the soil, water, air and health of human population. Extensive scientific researches have provided sufficient knowledge regarding the usefulness of earthworms (Levi-Minzi *et al.*, 1986). The role of earthworm in the process of vermicomposting of wastes is a physical and chemical action. The physical process includes substrate aeration, mixing as well as actual grinding while the biochemical process is influenced by microbial decomposition of substrate in the intestine of earthworm (Hand *et al.*, 1988; Arancon *et al.*, 2005). Vermicomposting through various wastes shows and gives chelating and phytoharmonal element in high content (Tomati *et al.*, 1995).

The vermicomposting process increase the mineralization rate when organic wastes passes through the gut of worm the nutrients converted from unavailable form to available forms, which consequently enrich the worm cast with higher quality plant nutrient (Gupta and Garg, 2007). During the vermicomposting there is a mineralization and stabilization of organic wastes substrate material with increasing proportion of C:N ratio and calcium enhancement (Garg *et al.*, 2005). However, vermicomposting process is generally characterized by measurement of the growth and reproduction of earthworm or determining different physiochemical parameters, such as changes in C:N ratio, variation of available nutrients and increases in humic substances (Suthar, 2006a). Suthar (2007b) reported that vermicomposting of crop residues and cattle shed wastes can not only produce a value added product but at the same time act as best culture medium for large scale production of earthworm. The vermicomposting of mixed vegetable wastes using cow dung influenced the rate of vermiconversion and increased the amount of macronutrients in the vermicomposts (Muthukumaravel *et al.*, 2008).

Thompsons and Nogales (1999) suggested that vermicomposting to ensure more favorable nitrogen mineralization–immobilization in the soil. Organic wastes accelerate the stabilization and increase the humic substance of organic matter in the vermicomposting (Fedrickson *et al.*, 1997; Neuhouser *et al.*, 1988). Brady and Weil (2002) and Welki and Parkinson (2003) reported that by hydrolyzation process the nitrogen was released in the form of ammonium ($NH_4^+$) with the help of micro-organism and it returned in to organic form). Szczech *et al.* (2001) reported that mineralization of nitrogen becomes slow in the latter phase of vermicomposting, resulting in increasing nitrate concentration. The earthworms increased the available forms of nitrogen and phosphorus in soil, increased metal bioavailability, and raised metal uptake in to plants by 16 to 53 per cent which the addition of exogenous humic acid to soil (Parr, *et al.*, 2002; Halim *et al.*, 2003).

Vermicomposting is a simple eco-biotechnological process in which earthworms are used to convert organic matter in humus like material called vermicompost. However, vermicomposting is a stabilization of organic material involving the joint action of earthworms and micro-organisms (Benitez *et al.*, 2000; Aira *et al.*, 2002; Benitez *et al.*, 2005; Suthar, 2006 a).

# 4

# Vermiculture

Large scale commercial production of earthworm's species for the benefit of soil and man in selected place or container adopting semi natural condition is known as vermiculture. Different species of earthworm have different potential with respect to the composting ability. Some species could be use for the preparation of high protein feed for aquaculture. Dead earthworm (*Eisenia fetida*) powder contains 62-64 per cent protein, 4.3 per cent lysine, 2.3 per cent cystine and 2.2 per cent methionin, 14 per cent fat 14 per cent carbohydrate and amino acids (Balaji, 1994; Ismail, 1997; Edwards and Bohlen, 1996). Vermiculture is a mixed culture containing soil bacteria and effective strain of earthworm (Ghosh, 2004; Board, 2008)

Studies have revealed that earthworms could be used for the production of cheap animal protein, weed management, as indicator organism for disposal of wastes, decomposition and production of organic fertilizers, plant litter, vegetable wastes, industrial wastes from fisheries, poultry municipal solid wastes and ecological soil management (Kaviraj and Sharma, 2003; Maboeta and Van-Rensburg, 2003).

Earthworms are the member of the phylum Annelida. The body is cylindrical with more or less uniformly placed ring annuli along the length of body. The dorsal surface of the body is recognized by its darker colour and a dark mid dorsal line which is due to the thick dorsal blood vessel situated just beneath the semitransparent skin. The ventral surface is marked by genital aperture and papillae located in anterior region of body. The eye, ear and lungs are absent. They breathe air that is present between soil particles diffuse through thin skins and they are forced through the surface if these are packed with rain water (Bhatnagar, and Palta, 1998). The worms are hermaphrodites *i.e.* both the sex organ are present in single individual but they requires another individual to mate because of the sexes are present at different segment. The clitellum that surrounds immature breeding earthworms secretes mucus after mating sperm from other worms is stored in sacs. As the mucus slides over the worm it encases the sperm and eggs inside. After slipping free from the worm, both end seal, forming a lemon shape cocoon approximately 1/8 inch long (Martin *et al.,* 1976).

There are about 3000 species of earthworm distributed all over the world (Gupta, 2005; Kale, 1998; Bhatnagar and Palta, 1998; Talashikhar and Dosani, 2008). Only half dozen species are beneficial to cultivation. The worms are detritivores as well as omnivore's animal but often selective in their food habit. They obtained their food from various organic matters, living bacteria, fungi, diatoms algae protozoan, nematodes and decomposing animals (Crossley *et al.,* 1971). They completely decomposed the wastes and stabilized it in to organic form. The worms are nocturnal and prefer the soil having sufficient moisture content, temperature and full of organic decaying matter (Barley, 1961). Aristotal referred to earthworms as "intestine of earth" (Kale, 1991). Darwins, (1881) treatise, the formation of vegetable mould through the action of earthworm with observation on their habits is regarded as the first modern scientific study of the beneficial role of earthworms and their casting in the soil (Bouche, 1977).

Kale and Krishnamoorthy, (1979) reported that the distribution of earthworms in soil influenced by several factor like soil texture, aeration, temperature, moisture, pH, inorganic salt, organic matter, dung, litter, reproductive potential and dispersive power. The soil of temperate region has higher percentage of earthworms but in tropical countries the humus feeder predominates over organic feeder worms. Kale and Bano (1988) reported that the external biotic parameters and poor nutritive resources are the major limiting factor for the earthworm's populations. The earthworms prefer the food rich in nitrogen content which increase growth and power of reproduction (Ranganathan, 2006; Dhavan *et al.,* 1991).

Generally the earthworms are called as bio-indicator of soil fertility. Earthworms support the healthy population of bacteria, fungi, actinomycetes, protozoan, insects, spiders, millipedes for sustaining a healthy soil (Rao, 1994; Ansari, 2008; Nelson and Oades, 1998). The role of earthworm in soil formulation and soil fertility is well documented and recognized (Ismail, 1995; Kale, 1998; Edwards *et al.,* 1995; Lalitha *et al.,* 2000; Shuster *et al.,* 2000). Soil fertility result from maintenance of soil condition. Decomposition process operates at a level adequate to release plant nutrients that will sustain optimum growth (Edwards and Lofly, 1972). Bhawalkar, (1990) has reported that the earthworms are wonderful bioreactor carrying out various functions in the soil. They take organic wastes and process it with gut micro flora and excrete the vermicastings, which is the effective biofertilizer for the crops. The earthworms are isothermal bioreactor, all the biological processes are sensitive to high temperature therefore they have novel temperature regulating mechanism (Kale, 1998; Gupta, 2005; Edwards and Baxter, 1992).

The earthworms provide a tremendous surface area for the microbial attack on feacal matter by mastication of organic waste through worm's gizzards. They also maintain stable pH of soil through their guts which contain various enzyme such as proteases, lipase, amylase, cellulase, lichenase and chitinase (Satchell, 1963). All enzymes are very active in a narrow range of pH (6.3 to 7.3).Earthworms can separate oxygen from air and supply the gut micro-flora thus encouraging various aerobic waste stabilization processes and destroying soil pathogens. They build up N-fixing activity in the soil by providing ideal condition of food moisture and air to N- fixing bacteria (Aira *et al.,* 2002; Gupta, 2005; Ranganathan, 2006; Talashikhar and Dosani, 2008).

In general earthworms are highly resistant to many pesticides and heavy metals. Worms could overcome the effect by increasing mucus secretion, restricting the movement and increasing the reproductive potential up to certain concentration levels (Senapati and Dash, 1984). Earthworms are also known to concentrate the pesticides and heavy metals in their tissues. Sodium was concentrated by *Eisenia fetida* but no lead (Bhole, 1992; Kaushik and Garg, 2003; Gupta *et al.*, 2005). The bioaccumulation of toxicants in the tissues of earthworms varies depending on the soil properties, pH, and calcium concentration (Govindan, 1998; Bhartiya and Singh, 2011, 2012a, b).

Generally the earthworms are classified in to epigeic, endogeic and anecic. The epigeic species live in the soil surface of 3 to 10 cm, and feeds on organic matter like leaf litter or animal excreta and decomposed it to manure containing necessary plant nutrients. These worms are useful for organic farming due to their high rate of fecundity and doubling power capacity. The endogeic earthworms live deep in the soil bellow than 10 cm and feed on humic material. The epigeic in particularly and aneceic in general have largely harnessed for used in vermicomposting process. Epigeic like *Eisenia fetida* and *Eudrillus eugenae* have been used in vermicomposting (Ismail, 1994; Senapati *et al.*, 1980). Surface dwellers are capable of working hard on litter layer and convert all organic wastes in to manure (Kale and Bano, 1988; Hartenstein *et al.*, 1979) The anacics, however, are capable of both organic wastes consumption as well as modifying the structure of soil, such burrowing species are widely used in soil management like the earthworm *Lampito marutii* (Ismail, 1993). These worms are not useful for soil cultivation because of very long life cycle and limited doubling capacity. The anecic worms can go very deep in to soil up to 60 to 90 cm and form complicated burrows for their movement and the mucus secretion. Some species of earthworms are used in organic farming *viz. Eisenia fetida, Perionyx excavatus, Perionyx sansibaricus, Lumbricus rubellus, Lempito mautri, Eudrillus eugenie, Dendrobeno veneta* etc. Among the vast diversity of earthworms available in India the *Allolobophora parvus, Pheretimas anomala, Octochaetona surnivsis* etc holds the potential for vermicomposting. However, different species of earthworms are used for management of different type of waste (Gupta, 2005; Ranganathan, 2006).

Suthar (2007, a, b, c) studied the decomposition efficiency of *Perionyx excavatus* and *Peryonyx sansibaricus* (Perrier) for the vermicomposting. Vermicomposting through earthworms is an eco-biotechnical process that transforms energy rich and complex organic substance in to stabilized humus like product (Benitez *et al.*, 2000). The microbes are responsible for biochemical degradation of organic matter, earthworms are the important drivers of the process, conditioning the substrate and altering the biological activity (Aira *et al.*, 2002). Beetz, (1999) and Benitez *et al.* (1999) reported that in vermicomposting process, inoculated earthworm maintain aerobic condition in the organic wastes, convert a portion of organic materials in to worm biomass and respiration product; and expel the remaining partially stabilized products vermicompost.

Several epigeic earthworms have been identified as detritus feeder which can be reared in organic wastes (Taylor *et al.*, 2003). Some species of earthworms of *Eisenia fetida, Perionyx excavatus* and *Eudrilus eugeniae* have appeared as the key candidates for organic waste recycling industries (Kale, 1998; Suthar, 2007a, b, c; Butt *et al.*, 1992; Butt *et al.*, 1995; Loh *et al.*, 2005; Garg *et al.*, 2006b). Earthworm accelerates the transformation of organic waste material in to stabilized forms by aeration and bioconversion, by their excreta and qualitative or quantitative influence upon the telluric micro flora (Suthar, 2007a; Vinceslas-Akpa and Loquest, 1997).

Gnanasoundari (1992) used the lampito mauritii for treatment of activated sludge for avoiding the environmental contamination, and suggested that the species prefer the place where C/N ratio becomes high. Kale and Bano (1991) have reported that time and space directly affects the growth and population of Eudrillus eugeniae. The earthworms are important biotic component of soil which makes soil healthy and check the environmental legislation by decomposing the decaying organic matter (Atlavinyte *et al.*, 1968). Due to rich amount of nutrients present in earthworm they are used as a food material for the cat fish Mystus vitatus (Arunachalam and Palanichamy 1984). Gajalakshmi *et al.* (2002a) have demonstrated the use of *Lampito mauritii* for vermicomposting of paper mill sludge. Gajalakshmi *et al.* (2001, 2002b) have suggested the use of *Eudrilus eugeniae in* vermiconversion of water hyacinth at different reactor. Only water hyacinth as a feed was not preferred by Eudrilus eugeniae, where as addition of cow dung approximately 14 per cent had positive impact on biomass growth and hatchlings production (Gajalakshmi *et al.*, 2005). Suthar (2007b) used the earthworm for management of crop residue with animal dung. Butt *et al.* (1992) showed that Lumbricus terrestris prefers on solid paper mill sludge under laboratory conditions. Solid paper mill sludge under laboratory condition has no deleterious effect on earthworms, although growth rate was poor. Elvira *et al.* (1996) studied the efficiency of *Eisenia andrei* (Bouche) in bioconversing paper pulp mill sludge. The mixture at 3:1 ratio was a suitable medium for optimum growth and reproduction of worm. Suthar, (2008, b) has studied that the bioconversion of post harvested crop residue with cattle shed manures by using earthworms *Eudrilus eugeniae.*

In India several exotic epigeic species like *Eudrilus eugeniae* and *Perionyx excavatus* are being used for vermicomposting (Ashok, 1994). Kaviraj and Sharma, (2003) have reported the comparative study of vermicomposting of municipal solid waste management through exotic species *i.e. Eisenia fetida* and local species *Lampito mauritii.*

Since the earthworms are rich in protein (65 per cent), fat (14 per cent), carbohydrate (41 per cent), and ash (3 per cent), therefore they are used as feed component for fish prawn shrimps (Jayashankar, 1994; Talashilkar and Dosani, 2008). They are also used as live stock feed in poultry industries. Earthworms have been staple food of some tribal people in Africa and New Guinea. In Philippines, earthworms are used as supplement for certain food dishes such as adobo and dinguan due to their high protein content.

# 5

# Vermicompost

Vermicompost is peat like materials with high porosity, aeration, drainage and water holding capacity (Ranganathan, 2006; Bhartiya *et al.,* 2011). Vermicompost have higher level of organic matter, organic carbon, total available NPK, micronutrients, microbial enzymes and plant growth regulators (Parthasarthi and Rangnathan, 2002; Chaioui *et al.,* 2003). Albanell *et al.* (1988) reported that vermicompost have less soluble salts, neutral pH, greater cation exchange capacity and humic acid content. Maheswarappa *et al.* (1999) reported that the vermicompost affects the soil pH and microbial population. Organic manures can increase soil organic matter, soil water retention and increased the aggregation of important organic compound (Zebrath *et al.,* 1999).

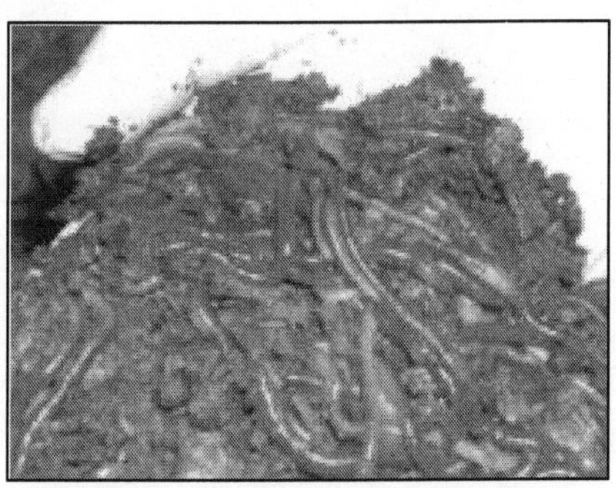

**Figure 1**. Vermicomposting by earthworm *Eisenia fetida*.

Venkatesh and Eevera (2008) reported that the concentration of macro and micro nutrient was found to increase in the earthworm- treated series of fly ash and cow dung combinations compared with the fly ash alone. Among different combination of fly ash and cow dung, nutrient availability was significantly higher in the 1:3 fly ash to cow dung treatment (Bhattacharya and Chattopadhyay, 2004; Venkatesh and Eevera, 2008). The vermicomposts treatment increased scorodose accumulation, which was directly related to the harvest index, resulting in greater yield and bulb quality of garlic (*Allium sativum* L.) (Ledesma *et al.*, 2000; Arguello *et al.*, 2006). Ansari (2007) reported that application of vermicomposts in sodic soil improve the physical, chemical and biological properties of the soil.

Manivanan (2004) and Ramamoorthy (2004) have shown the water holding capacity of clay loam soil and sandy loam soil to have enhanced when vermicast was added at the rate of 5 tonnes/hact. The cation exchange capacity of vermicomposts has the potential to correct the acidity and alkalinity of soil as well as improving the physical properties of the soil (Venkatesh and Eevera, 2008). Ramamoorthy (2004) has clearly shown that vermicompost keeps the natural pH of clay soil and sandy soil, whereas, application of NPK to soil tends to make slightly alkaline.

Vermicomposts is a homogenous, retain most of the original nutrients and has reduce levels of organic contaminants with respect to initial feed mixture because they are degraded (Ndegwa *et al.*, 2000).Vermicomposts, used as soil additives, or as a components of greenhouse bedding plant container media, have improve seed germination, enhanced seedling growth and development, and increased productivity (Atiyeh, *et al.*, 2000). Various kinds of vermicomposts accelerates organic matter stabilization and gives chelating and phytohormonal elements which have a high content of microbial matter and stabilized humic substances (Arancon *et al.*, 2005).

The positive influence of earthworm on the living nature can not compared with other because they converted organic matter into humus peat which provides soil fertility and safety for soil (Gupta, 2005). Atiyeh *et al.* (2000) found that vermicompost have higher nitrate content, which is the more plants available form of nitrogen. Hammermiester *et al.* (2004) indicated that vermicomposted manure have higher nitrogen availability than conventional compost. Fulekar and Jadia (2008) reported that plant litter contained more available phosphorus after ingestion by earthworms and they attributed to physical breakdown of the plant material. Davis *et al.*, (2002) reported that an average phosphorus contents for dairy manure compost is 22 lbs $P_2O_5$ per tones and also find out nutrients variability in this manure.

Sharma *et al.* (2011) observed that vermicompost of spinach wastes have significant higher contents of sodium, magnesium, phosphate and potassium with respect to their compost. Vermicompost of tannery sludge mixed with cattle dung shows significant increase in the level of nitrogen, sodium, phosphorus, and pH whereas decrease in potassium, organic carbon and electrical conductivity (Vig *et al.*, 2011).

# 6

# Vermiwash

Apart from vermimeal or vermiprotein, vermiwash is the organic fertilizer decoction obtained from units of vermiculture/vermicomposting as drainage. It is used both as foliar spray and in the root - zone of the plants. No special device is requiredto collect the vermiwash except for a tap tobe put up at the bottom of the containers where earthworms are cultured. The enzymes present in the extraction stimulate the growth and even develop resistance in crops receiving the spray.

Vermiwash is the liquid extract collected after the passage of water through the different layers of worm culture unit. The extract contains excretory products of earthworms, secritions of the worm, the coelomic fluid oozing through the dorsal pores, mucus, enzymes secreted by the worms and the microorganisms, plant

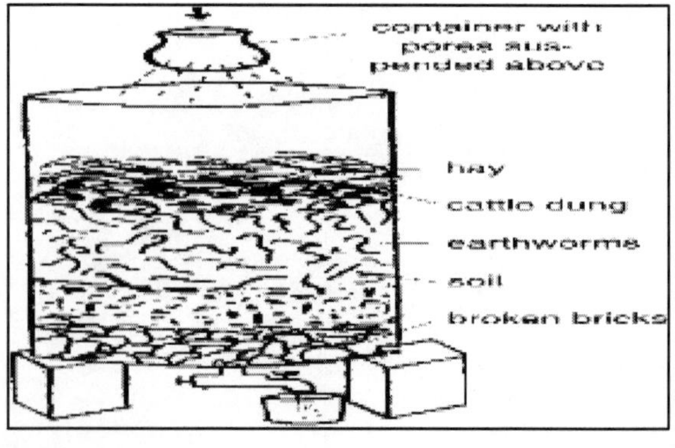


**Figure 2**: Vermiwash Collecting Device.

nutrients, vitamins and plant growth promoting substances. The water also carries with it all the dissolved substances resulting out of the washing of the live and dead earthworms, soil micro-organisms and decomposing organic matter. As the water could dissolve some vermicast, which contain lot of nutrients, particularly K, Ca and Mg, they find their way into vermiwash. Vermiwash was found to be develop resistance to diseases in plants. It was found to be beneficial in nurseries, lawn and orchards.

Vermiwash is the extract of vermicomposts containing rich amount of earthworms. It contains micronutrient, vitamins, hormones and disease resistance power (Grappelli *et al.*, 1987). Vermiwash is honey brown in color having heterotrophic bacteria, fungi, actinomycetes, including nitrogen fixer phosphate solbuliser and enrich with macro, micro nutrients, enzyme hormone and vitamins (Lozek and Fecenko, 1998). It is liquid organic biofertilizers, pesticidal in nature (Kale, 1998; Grappelli *et al.*, 1987; Umamaheswari *et al.*, 2003). Vermiwash, liquid manure is very useful as a foliar spray to enhance the plant growth, yield and to check the development of disease. The vermiwash complex is efficient in rising of nurseries, lawns and orchids (Ismail, 1995; Pramoth, 1995). Giraddi (2001) have studied that wash of earthworm is a plant promoter substance.

Vermiwash is the coelomic fluid extraction; it contains several enzymes, plants hormones like auxines, cytokinin, gibberellins and vitamins, especially $B_{12}$ along with micro and nutrients. This liquid manure is collected in the liquid form and used as foliar spray, which stimulate the growth and yield of crops (Ismail, 1997). It also increases the resistance of crop against harmful disease (Shields and Earl, 1982; Shivsubramanian and Ganeshkumar, 2004). Gamaley *et al.* (2006) have suggested that vermiwash is foliar manure root nutrition and optimized the productivity of crops. Vermiwash a best tonic for plant show significant growth and productivity of crops (Ismail, 1997). This is attributed to better growth of plants and higher yield by slow release of nutrient for absorption with additional nutrient like gibberellins, cytokinin and auxins by the application of organic vermicomposts with vermiwash (Subler, 1995; Raviv *et al.*, 1998).

Earthworm produced bacteriostatic substance found in the vermiwash which protect the plant from bacterial infections (Pathak and Ram 2004). Weerasinghe *et al.* (2006) has reported that vermiwash is the wash of earthworm's ceolomic fluid. Calcareous layer and the watery extract of bedding materials which contains soluble micro and macro nutrients, natural plants growth hormones beneficial microbes vitamins and amino acids. It also occurred pesticidal properties. Vermiwash are recognized for foliar spray in morning before sunshine and evening, after sun set and also suggested that cow urine and vermiwash the ratio of 1:1 diluted by ten time of water is an effect biopesticides and liquid manure (Subasahri, 2004; Ismail, 1997; Pramoth, 1995).

Zaller (2006) has suggested that various effect of foliar spraying of vermicomposts extract on fruit quality and indication of 'Late blight' suppression of field grown tomatoes. Hoffland *et al.* (2000) have reported that nitrogen concentration of tissue is significantly altered by spraying of vermiwash. It also suppresses the disease causing pathogens in tomato plant. The application of aqueous compost extract has been shown to reduced the necrotrophs as well as biotrophs elocote disease (Weltzien, 1989; Fokkema, 1993; Al- Dahmani *et al.,* 2003). Aqueous extract of vermicomposts have shown to depress the soil born pathogens and pests (Orlikowski, 1999; Szczech *et al.,* 1993; Nakasone *et al.,* 1999; Rodriguez *et al.,* 2000). Since composts and vermicomposts extract contain a high amount of nutrients, it is reasonable that these extract could also be used as foliar fertilizers. Generally, foliar spray would offer a method of supplying nutrients to higher plants more rapidly than root application (Marschner, 1995; Shweta *et al.,* 2004).

Under dry condition foliar application of nutrients is much more effective than soil application (Grudon, 1980). It has been reported that spraying of vermiwash on variety of tomato caused significant increase in the growth of plant and yield of fruits (Zaller, 2006; Siminis *et al.,* 1998; Atiyeh *et al.,* 2000; Arancon *et al.,* 2003, 2005). Foliar sprays containing nutrient can also compensate the decline in nutrient uptake by roots with the onset of the reproductive stage as a results of sink competition for carbohydrates (Trobisch and Schilling, 1970). Intelligent and selective use of organic amendment like vermicomposts, vermiwash, mulch (chiefly including plant residue like paddy) and green manure have effective in soil conditioning and soil property (Ansari, 2007; 2008).

Effect of added liquid from pots containing earthworms have been shown to increase dry matter production of rye-grass (Graff and Markeschin, 1980). Latter it was demonstrated the growth of ornamental plant after adding aqueous extract from vermicomposts shows similar growth pattern as with the addition of auxins, gibberellins and cytokinins through the soil (Tomati and Galli, 1995; Grappelli *et al.*, 1987; Tomati *et al.*, 1987). Conventional thermophilic composts microbial population or humic substances present in the vermiwash can influence the fruit quality of plant (Edwards *et al.*, 2004). The presence of large number of beneficial microorganism present in vermiwash help in plant growth and protect it from a number of infections. Edwards *et al.* (2004) discussed the hormones produced by vermiwash are very effective for plant growth and its diseases suppression. The use of vermiwash in leaf areas of plant suppresses the plant parasitic nematode and arthropods pest and improving the growth, productivity and seed germination of plants (Zaller, 2006). Increased microbial activity in vermiwash results in the production of significant quantity of plant growth regulators such as indol acetic acid, gibberellins, cytokinins (Edwards, 1998; Krishnamoorthy and Vajranabhiah, 1986). Large quantity of humic acids was produced during vermicomposting, which leaches out from vermicomposts during extraction of vermiwash (Ismail, 1997). Humic acid have positive effect on plant growth (Manivannan, 2004, Ramamoorthy, 2004; Atiyeh *et al.*, 2002). Extract of thermophillic composts proved to be effective against various fungal diseases of leaves and fruit (Scheuerell and Mahaffee, 2002).

Tripathi and Bharadwaj (2004) has reported that the nitrogen in the form of mucus, nitrogenous excretory substances, growth, stimulating hormones and enzyme in vermiwash play significant role in germination of seed and development of seedlings of legumes. The significant growth was observed in the black gram spraying by vermiwash (Sudha *et al.*, 2003). Viveka *et al.* (2005) reported that vermicomposted weeds and its aqueous extract significantly affect the growth and productivity of okra plant. Vermiwash is a natural plant growth supplements for tea, coconut and horticulture crops (Weerasinghe *et al.*, 2006). Vermiwash is a organic source of fertilizer having inorganic N and K. The foliar application of vermiwash promotes the physiology of plant thereby an increase in the yield and quality of product. Significant increase in plant height, number of leaves per plant and chlorophyll content was observed differ 10 per cent diluted foliar spray of vermiwash (Pathak and Ram, 2004).

Kobatke (1954) reported that the coelomic fluid from earthworm body had antibacterial property and its foliar spray on vegetables increased the quality and quantity of yield. It was also observed that foliage turned dense green in to two three days when spray was used on plant other than vegetables (Anonymous, 1993). Ismail (1997) reported that vermiwash can be sprayed on plant as a foliar spray for improving of quality and yields of okra crops.

Snieg and Bury (1998) studied the influence of a water extract of vermicomposts on growth and development of potatoes crop. Lozek and Gravova (1999) have observed that tomato seedling planted in medium heavy black soil in the green house when supplied with vermiwash resulted increased 7.3 per cent yields and decreased in fruit nitrate content by 50 per cent. Todkari (2001) studied the effect of vermiwash on growth characteristics, yield of plants and inferred that vermiwash have good nutrient potential. Maximization of yield of flowers, like chrysanthemum and marigold is fertilized with foliar spray of 100 per cent vermiwash, indicate a quick absorption of the nutrients through foliage for better nourishment of beach flowering plants (Todkari and Talashilkar, 2001).

The significant bio-pesticidal properties of vermiwash prepared form cow dung and vegetable wastes were observed against powdery mildue disease of cow pea (Balam, 2000; Subasashri, 2004). The vermiwash contain enzyme cocktail of proteases, amylases, ureases and phosphatase and also reported that vermiwash contain nitrogen fixing bacteria like azobacter, agro bacterium and rizobium species and some phosphate solublizing bacteria (Chaudhary, 2005; Zambare *et al.*, 2008). The microbes present in the vermiwash significantly influenced the biological cycle of phosphorus, present in organic compounds. The bio-geochemical cycles of phosphorus are decomposed and mineralized by enzymatic complex like phosphatase produced by microbes (Chaudhary, 2005; Zambare *et al.*, 2008). Desai (2003) have conducted a field experiment the effect of city compost and seavage sludge with and without vermiwash on the growth parameter, dry matter and flower yield and found that vermiwash is suitable for quick absorption of major nutrient and better nourishment of china aster.

Vermiwash is a pale yellow liquid biofertilizers. It is a mixture of excretory products and mucus section of earthworms (*Lampito mauritii* and *Eisenia fetida*) and potent fertilizer for better growth and yield of plants (Yadav *et al.*, 2005). Vermiwash enhanced the growth and yield of soyabean (*Glycine max*) by foliar application (Rana, 2000). Karuna *et al.* (1999) studied the stimulatory effect of vermiwash on crinkle red variety of Anthurium and Reanum. Vermiwash caused significant increase in growth and productivity in marigold plant (Shivsubramanian and Ganeshkumar, 2004). George *et al.* (2007) demonstrated that vermiwash spray significantly maximized dry chilly yield.

Weekly application of vermiwash increased the radish yield by 7.3 per cent (Buckerfield *et al.*, 1999). Thangavel *et al.* (2003) have observed a significant increase in the yield of paddy after application of vermiwash. Thangavel *et al.* (2003) also reported that the vermiwash affect the soil nutrient as well as growth productivity of crop. It was reported to have yielded good results especially early initiating and long lasting inflorescence of Anthuriums (Rao, 2005). It could also be used as liquid fertilizer (20 to 30 per cent dilution) inhibits the mycelial growth of pathogenic fungi (Rao, 2005). Buckerfield *et al.* (1999) have shown that vermiwash leaches from vermicompost in active worm beds enhanced the growth of radish when applied at seedlings.

# 7

# Biopesticides

Since the practice of agriculture began, humans have been struggling to reduce the adverse effects of pests on crops, forests and other human managed ecosystem. Pests have been are and will still continue to be a major restriction to agricultural production throughout the world. Man has been fighting against his pest enemies from the day he learnt the art of agriculture. Increase in food production is necessary to fullfill the current demand of increasing population, for which protection of crops from pest is a basic matter for concern. Due to pesticides role in modern management practices, their use cannot avoid in agriculture. It has been estimated that crop damage is high as 20 per cent where pesticides are not used. Decrease in crop losses by pests and disease will be an important factor achieving the target of productivity and this target can be hold by the help of pesticides. The use of pesticides helped in eradication of diseases and also in boosting crop production. Their elimination is essential in the interest of food production and also public health. In the modern time, the use of synthetic pesticides has become a necessary in various branches of our economy. Fast expanding human establishments, intensive agriculture and higher input rate of waste material in to the environment have created additional resources for pests and other harmful pathogens to multiply. The application of synthetic pesticides seems to be the only effective solution, but most pesticides are dangerous due to their poisonous nature. Plants sprayed with pesticides become toxic and contaminatin the soil due to the fall on the ground. The pesticides are detoxified in the soil by adsorption and degradation. However, their presence in the soil for a long time could adversely affect its fertility, besides polluting nearby waterbodies. The presence of pesticides in ground water is a matter of serious concern because pesticides cannot be filtered out by the use of modern filtration techniques and contaminated ground water can be main source of pollution for crops if it is used for irrigation purposes. As these chemicals move into soil, they create many problems. Few pesticides are not readily biodegradable and persists in the soil or in water for many years. Others are detrimental to non - target organisms such as beneficial insects and some soil organisms. The selective action of pesticides is never perfect and many target organisms are effected by their toxicity, some of which may be useful organisms. The elimination of some insects and bees which vitally aid in pollination of many plants could cause considerable damage to agricultural productivity. It is difficult to prevent the circulation of these chemicals in the environment due to their persistence. Thus, once applied they continue to harm the non-target organisms for long periods of time. They are bioaccumulated and biomagnified, which cause problems at higher tropic levels in an ecosystem.

The plant extracts/products have attracted the attention world over due to various advantages over the conventional pesticides, such as elimination of residue hazards, contamination, pollution, biodegradation and lower cost. Of late interest has renewed in the pest management potential of natural products. These products are the compounds that have evolved in plants for defence against phytophagous insects. The products of plant origin have been investigated for antifeedents/feeding deterrent activity, ovicidal and growth inhibitory action thus acting as potential source for insect control. It is interesting to mention that plants evolved potent chemical defences, which prevent insects from feeding and thus are useful in pest management. Biopesticides are inherently less harmful then conventional pesticides. They are designed to affect only one specific pest and often are effective in very small quantities and often decompose quickly. When used as a component of Integrated Pest Management (IPM) programme, biopesticides, can greatly decrease the use of conventional pesticides, while crop yield remain high. Biopesticides in general have a narrow target range and a very specific mode of action and slow acting. They suppress a pest population and have limited field persistence and a short self life. They are safer to humans and the environment than conventional pesticides and present no residue problems.

Biopesticides are of mainly three types, *viz.*, Microbial pesticides, Plant pesticides and Biochemical pesticides. Microbial pesticides contain a microorganism (bacterium, fungi, virus, protozoan or algae) as the active ingradients. Most widely used microbial pesticides are *Bacillus thuringiensis* (*Bt*) which can control certain insects in cabbage, potatoes and other crops. Bt produces a protein that is harmful to specific insect pests. Plant pesticides are substances that plant produce from genetic material that has been added to the plant. Biochemical pesticides are naturally substances that control pests by non-toxic mechanisms. Biochemical pesticides include substances that interfare with growth or mating, such as plant growth regulators or substances that repel or attract pests, such as pheromonones.

Plants can be insecticidal, antimicrobial to bacteria, fungi, and viruses. Some are also herbicidal some possass other type if biological activities. The future of botanicals in the modern concept of integrated pest management system appears to be very promising as they are biodegradable in nature and much safer to higher animals including humansbecause of their low mammalian toxicity. These beneficial, bioactive chemical substances are found in abundance in plant species. The combination of biopesticides with vermiwash and vermicompost minimized the infestation rate of different pest of different crops.

## *Azadirachta indica* A. Juss (Neem)

Neem is hailed as wonder tree, kalpavriksha and miraculous tree for its versatile use and has recently come to the focus of global attention. In fact, this tree plays a great role of village dispensory for its therapeutic efficiencies. Seed kernals produce a dark brownish- yellow, bitter oil. The goodness of neem lies in active chemicals "Limnoids" which have been isolated from oil, seeds, bark and leaves of neem. The most important limnoid, azadiractin A and other similar compounds such as salanin, melinatriol are found extremely effictive against insects even in minute quantities. Neem constituents through diversified biocidal activities, offer a novel approach in pest management. Some commercially avaible neem based pesticides products in india are Bioneem, field marshal, Godrej achook, Jawan crop protector, Margocide, Neemark, Neemgold, Neemol etc. Azadiractin acts as a strong antifeedent to locust at 40 micro-gram per litre concentration. Similarly, melintriol, a compound isolated from neem seed oil, is a prominent locust antifeedent.

## *Allium sativum* L.

Garlic (*Allium sativum* L.), is a valuable spice plant used as a food item as well as medicine in different parts of the world. At the beginning of the present century garlic was used in medicine on the basis of traditional experience passed from generation to generation. Garlic has been used for medicinal purposes for centuries. Scientific evidence established that garlic does indeed possess antimicrobial or medicinal properties (Cavallito and Bailey, 1994). Recent researches indicate that garlic extract has anti-microbial activity against many genera of bacteria, fungi and viruses. These researches were the first to isolate the major active constituent from garlic bulb by steam distillation of ethanolic extracts. The compound identified as allicin, a diallyl thiosulphate (2-propenyl 2- popenethiole sulfinate). It was reported that allicin, at a concentration of 1:85,000 in broth was bactericidal to a wide range of gram - ve and gram +ve organisms. Intact garlic cloves contain only a few medicinally active compounds. The main chemical constituents of intact garlic is the amino acid alline an alkyl derivative of cysteine alkyl sulfhoxide, which may varies from 0.2 to 2.0 per cent fresh weight. The biological impact of garlic chiefly depends on the way of prepration of garlic extract. The crushing, chewing or cutting (or exposing dehydrated, pulverized garlic to water) of garlic cloves releases the vascular enzyme allinase that rapidly lyses to cytosolic cysteine sulfoxide to form sulfhenic acid (R-SOH). The sulfenic acid immediately condenses to form allicin, the compound produce the odour of fresh cut garlic. At least 100 sulpher containing compounds basic to medicinal uses of garlic are also found. Allicin represents 70-80 per cent of the total thiosulfinates formed. Garlic's role in preventing cardiovascular disease has been acclaimed by several research groups. Chemical constituents of garlic have been investigated for possible effect on hyperlipidemia, hypertention, platelet aggregation and blood fibrinolytic activity. Garlic besides its role as flavourings and seasoning, present strong medicinal, preservative and antioxidant properties and thus contribute to overall safety and preservative of food products. In the recent years, the growing demand for novel products that are safe, natural and fresh, is stimulating the research into new processing methods to improve microbiological quality of foods and beverages, with no marked changes in nutrient content and more sensory attributes. Recent researches in field of pest control indicate that garlic prepration has strong insecticidal, nematicidal, rodenticidal and molluscicidal activity. Although field trials and laboratory experiments on the pesticidal activity of garlic has been conducted, yet the more researches are recommended to study the exact mode of action, way of delivery in environment for effective control of pest.Allicin- an active nematicidal principal in garlic-has been isolated and tested against Meloidogyne incognita infesting tomato. Allicin at 25 ppm for 5min as a root-dip treatment for tomato seedlings is effective against *M. incognita.* Sing and Sing,(1993) studied on *Allium sativum* (garlic)and recommended it a potent insecticide. Singh andSingh, (1996) studied the enzyme inhibition by allicin which is the active insecticidal agent of garlic(Allium sativum). The presence of antibacterial activity in heated garlic extract in which allinase failed to act and to identify the inhibitory compounds in heated garlic. The isolation, purification, identification,synthesis and kinetics of activity of *Allium sativum* and a hypothesis for its mode of action. The chemistry of garlic and its insecticidal activity on the crops. The antimicrobial effects of *Allium sativum* L. (Garlic), *Allium ampeloprasum* L. (Elephant garlic) and *Allium cepa* L. (Onian), Garlic compounds and commercial garlic supplement products. The HPLC analysis of allicin and other thiosulfinates in garlic cloves and homogenates. The garlic does indeed possess antimicrobial or medicinal properties, the garlic juice was a very potent insecticidal agent.

# 8

# Effect of Vermicompost and Vermiwash on Plant Growth and Productivity

The vermicompost stimulate plant growth providing sufficient amount of nutrition (Atiyeh *et al.*, 2002; Arancon *et al.*, 2004). The vermicompost contain plant growth regulator and plant growth hormones that increase the growth and yield of different crops (Canellas *et al.*, 2002). Uses of vermicompost promote the soil segregation and stabilize the soil structures, improve the air water relationship with soil, thus increase the water holding capacity and enhance the root development of plants (Marinari, *et al.*, 2000; Gupta, 2005). Sreenivas *et al.* (2000) studied the integrated effect of vermicompost resulted the soil nitrogen level significantly increasing up to 50 per cent in gourd (*Luffa acutangula*). Jadhav *et al.* (1997) reported that nitrogen, phosphorus, potassium and magnesium were high in rice (*Oryza sativa)* plant attained highest height with respect to fertilizer when vermicompost were applied in the combination.

The application of vermicompost had not only enhanced the soil fertility but also increased the rice yield (Vasanthi and Kumarasamy, 1999). Thomsen (2001) reported that that using 226kg composted H15-labeled wheat straw and manure per hectare, approximately 65 per cent of compost nitrogen was assimilated by wheat (Triticum astivum) crop and 2.6 per cent was utilized by subsequent barley (Hordeum vulgare) crop. Ndegwa *et al.* (2000) observed that nitrogen increases between 14-39 per cent in soil after application of vermicomposts. Kumar and Singh (2001) reported that several strain of N fixing and phosphate solublizing bacteria in vermicomposting bed and founded that increases of available phosphorus and nitrogen level.

Desai *et al.* (1999) studied the effect of vermicompost of 10-20 per cent in the soil and observed that 40 per cent increase in yield of wheat. Agrwal *et al.* (2003) studied singly and in binary combinations of vermicomposts on the growth and yield of wheat (variety-HD.2643) and found that in all treatments increase the total biomass production and yields of wheat plant over control. Karmegam *et al.* (1999) studied that effect of vermicompost on yield of maize in different textural soils and their uptake of nutrient elements. Bawa (1995) reported about the efficiency of vermicompost with comparison to farmyard manure and city compost, on the groundnut crop. In the experiment, a positive response was obtained after application of vermicompost in to field crop like sorghum (Patil and Sheela, 2000) and sunflower (Devi and Agrwal, 1998). Ramamurthy (2006) reported that application of vermicompost improved the productivity and quality of Nagpur mandarin (Citrus reticulata blanco). The influence of organic manure on the growth and yield of tomato (Lycopersicum esculentum) and soil chemical properties with positive results (Arancon *et al.*, 2003; Talarposhti and Kambougia, 2007). Reganold *et al.* (2001) reported that after single application of vermicompost on cherries, the yield and quality consistently increases for next 3 years.

The yield of pea (*Pisum sativum*) was higher with the higher application of vermicompost along with recommended nitrogen, phosphorus and potassium in comparison with alone of these fertilizers (Karmegam and Danial, 2000). Vadiraj *et al.* (1998) reported that the yield of coriender obtain were positive then chemical fertilizer. Nethra *et al.* (1999) reported that fresh flower weight, number of flower (26 per plant), flower diameter

(6cm/flower) and yield (0.5 tonnes/ha) of Chrysanthemum chinesis increased with application of different level of vermicompost. An application of vermicompost at 10 tonnes/ha along with the recommended dose of P and N resulted into 50 per cent increase in dry pod yield than the recommended dose of nitrogen and phosphorus alone (Nethra *et al.*, 1999).

Vermicompost had been shown to influence the growth and productivity of a variety of plants, cereals and legumes (Chan and Griffith, 1988), vegetables (Atiyeh *et al.*, 2000), ornamental and flowering plants and field crops (Fulekar and Jadia, 2008). Vermicompost improved seed germination, enhanced seedling growth and productivity (Atiyeh *et al.*, 2001). Large amounts of humic acid were produced during vermicomposting and reported to have positive effects on plant growth (Manivannan, 2004). Canellas *et al.* (2002) demonstrated that there were exchangeable auxin and humic acids extracted from vermicompost which enhanced the growth of plant.

Sagar *et al.* (2002) compared the growth of Ocimum sanctum (an oil crop) in vermicompost of cow manure, farmyard manure, urea and control medium and they observed that oil yield, leaf weight and plant weight was 25 to 52 per cent higher in vermicomposts treatment than control. Ozores-Hampton and Vavrina (2002) observed that growth of tomato increased linearly with vermicompost content, from 0.08 to 0.22g dry matters per plant where as shoot biomass of control was 0.11g dry matters per plant. Vermicompost applications resulted in a 31 per cent increase in productivity whereas reduction of harmful leaf chlorine content from 10 to 4 mg/g of dry matter (Rajkhowa *et al.*, 2000). Klock-Moore (2001) used farm yard manure and biosolid compost in a field; compost has proven successful as a fertility source when applied at heavy rates in to crops field. Yermiyahu *et al.* (2001) observed that the reductions of boron (B) uptake by pepper (Capsicum frutescens) as rates of cow manure and straw compost increased from 30 to 100g/kg compost soil.

Pascual *et al.* (1997) noted that increases of biomass and basal respiration of useful micro-organism in an arid soil amended with vermicompost of municipal solid wastes and sewage sludge. Suwan *et al.* (2006) observed significant increases of mycorrhizal population of bean (Phaseolus vulgaris) roots in coarso-silty topsoil amended with vermicompost. The beneficial microbes of vermicompost may reduce the resources of excudes from root and seed tissue of plant which is necessary for germination of pathogenic fungal spores (Stone, 2002). The presence of plant growth promoting rhizobacteria (*Bacillus* sp. and *Pseudomonas* sp.) leads to increases of chitinases, beta-1 and 3-glucanases, peroxidases, and other important pathogenesis-related proteins as well as the accumulation of antimicrobial low molecular weight substances and the formation of protective biopolymers (like-lignin, callose, glycoprotiens) (Chan *et al.*, 2000). Tuitert *et al.* (1998) reported the much suppression of *Rhizoctonia solani* in cucumbers (*Cucumis sativus*) planted in 20 per cent mature house-hold waste compost.

Bullock and Ristaino (2002) reported the southern blight (Sclerotium rolfsii) incidence was only 3 per cent in field tomatoes after amended of cotton gin trash composts. Godase and Patel (2001) used controlled microbial compost, he find that methyl bromide treatments exhibited 100 per cent root colonization by Rhizoctinia fragariae in next 2 year. The pure cow manure vermicompost inoculated in the field of phytopthora nicotianae, Fuasarium oxysporum and Plasmodiophora brassicae which infect the tomato seedling and clubroot disease in cabbage(Brassica olaraca capitata) by Phytopthora, Pplasmodiophora) was 75 to 300 per cent less during development. There has considerable evidence in recent years regarding with the ability of vermicompost to protect the plant against various diseases. Edwards and Arancon (2004) reported that the vermicompost applications suppressed the incidence of disease and repelling of pest significantly. The pest reduction from crop due to breakdown of chitin in the exoskeleton of insect by chitinase enzyme that present in vermicompost of earthworms (Arancon, *et al.*, 2004; Edwards and Arancon, 2004). Vermicompost could be used for destroying of pathogen successfully when used with biopesticides (Eastman *et al.*, 2001; Appelholf, 2003). Edwards and Arancon (2004) reported that vermicomposts useful for the reduction of arthropods (aphid, milybugs, spidermites) population and also subsequent reduction of plant damage (tomato, pepper and cabbage). Vermicompost can be considered as alternates of pesticide for pest control (Edwards and Arancon, 2004). There was significant effect of vermicompost of cattle dung on the growth and yield of tomato (Joshi and Vig, 2010).

Shukla and Singh (2013) reported that the vermicompost of different combination of animal dung, agricultural and kitchen waste have significant effect on the growth, flowering and productivity of certain crops the maximum growth was observed after 50 days in buffalo dung with gram bran for okra, lobia and bitter gourd among all combination of vermicompost. The significant early flowering period was recorded in cow dung with vegetable wastes for both okra and lobia while buffalo dung with rice bran for bitter gourd. The maximum productivity of okra, lobia and bitter gourd has been recorded in verimcompost prepared from cow dung with vegetable waste in the comparison to control. Nath and Singh (2012) also reported that the effect of combinations of vermiwash with biopesticides against aphids infestation and growth and yield of mustard.

The vermicompost of buffalo during with straw, horse dung with gram bran, combination of straw with buffalo dung have significant growth of paddy, maize and millet crops (Nath and Singh, 2012). Nath and Singh (2011) observed that the foliar spray of vermiwash with biopesticides caused a significant effect on growth, flowering and productivity of soyabean.

# 9

# Earthworms

Earthworms belonging to phylum- Annelida, class-Oligochaeta. The body of earthworm cylindrical with more or less uniformly placed ring annuli along with the length of the body. There are over 3000 species of earthworm but dozen of species are useful for vermicomposting (Ranganathan, 2006) Some species are efficiently ploughing the land and recycle the organic matter for growth of the plants. The distribution of earthworm in soil is influenced by several factors such as soils textures, aeration, temperature, moisture, pH, organic salts, organic matter, dung and reproductive potentials (Garg, *et al.*, 2006a; Suthar, 2006 a). The pH highly influenced on diapauses condition in earthworm (Aalok *et al.*, 2008). Majority of worms occurred in soil where the moisture ranges from 12-45 per cent (Govindan, 1998). The external biotic parameter and the nutritive resources of the soil are primary controlling factor for their population increase (Albanell *et al.*, 1988; Edwards and Bholen, 1996).

Generally earthworms are classified into as anecic, endozoic and epigeic. The anecic worms are burrowing that come to the surface at night to drag the food down into the permanent burrows deep with the mineral layer of soil (Arancon *et al.*, 2005). The endozoic worms are also burrowing but their burrows are much less, rarely come over surface from deep and feed on organic matter of the soil. Most of the endozoic species have very long life cycle with limited regenerative capacity (Ranganathan, 2006). The epigeic species worm live in the surface litter and feed on decaying organic matter. They are very active with high regenerative capacity within a short period of time (Gupta, 2005).

Earthworm and microbes symbiosis act as controlling system of enzyme during the metabolism and keep the content of nutrient in vermicast (Parthasarathi and Ranganathan, 2002). Five species of Earthworm *Eisenia fetida, Denderobaena vaneta, Lumbricus rubellus, Eudrilus eugeniae* and *Peryonix excavatus* have been recommended to use in breakdown of organic matter (Talashilkar and Dosani, 2008). In India two species are extensively used for vermiculture namely *Eisenia fetida* (redworm) and *Eudrilus eugineae* (night crawler) along with the exotic species like *Denderobaena vaneta, Peryonix excavatus,* and *Lumbricus rubellus* (Edwards *et al.,* 1995; Kaushik *et al.,* 2003). Scott (1988) observed that worm-digested animal wastes are as much supplement to peat in loam-less composts for horticulture. Repeated annual applications of poultry litter and animal wastes lead to increased nutrient metal concentration in soil. Ghosh (2004) reported that vermiculture is an innovative biotechnology, in which the breeding and propagation of earthworm *Eisenia fetida* and the use of its castings become an important tool of wastes recycling converting in to vermicomposts. Aalok *et al.* (2008) have studied the use of earthworm as natural bioreactors for cost-effective and environmentally appropriate waste management.

Ismail (1993) recommended that the species *Lempitto mauritii* and *Perionyx excavatus* are suitable for vermicomposting and soil management in South India. Ranganathan (2006) observed that *Denderobaena vaneta, Perionyx excavatus, Lumbricus rubellus* and *Perionyx sensibaricus* are suitable for the solid wastes management. Epigiec species *Eudrilus eugeniae* have been extensively used in converting organic wastes in the vermicomposts (Aalok *et al.,* 2008). Venkatesh and Eevira (2008) observed that the different combinations of fly ash and cow dung with inoculation of Eudrilus eugeniae, the nutrient availability was significantly higher in the 1:3 (fly ash to cow dung) treatments. Phosphate solublizing microbes such as *Micrococci, Pseudomonas* spp, *Bacillus* spp, and *Aspergillus* species were observed in gut and cast of *Lempitto maurotii, Perionyx excavatus* and *Eudrilus eugeniae* (Parthasarathi and Ranganathan, 2000). Mba (1997) reported that earthworm cast of Eudrilus eugeniae were rich in rock-phosphate solublizing microbes and have high phosphate solublizing capacity with nitrogen fixing cellulolytic microbes.

Mixture of gaur gum industrial waste (a ligno-cellulogic wastes of *Cyamposis tetragonaloba* L.), cow dung and saw dust (60:20:20 ratio) is an ideal combination for enhancing maximum biopotential of earthworms (*Perionyx excavatus*) for management wastes as well as for earthworm biomass and cocoon production (Suthar, 2006 b). Suthar (2007 d) reported that the vermicomposting efficiency of Perionyx excavatus was influenced by the different waste materials. The crop residues can be used as an efficient culture media for large scale production of *Eudrilus eugeniae* (Suthar, 2008 a). Suthar (2007b) demonstrated that quality of waste material like jowar, bajra straw and sheep dung manure in vermiculture influence the biomass and reproduction of *Eudrilus eugeniae, Perionyx excavatus* and *Parionyx sansibaricus*.

# 10

# *Eisenia fetida* (Savigny)

*Eisenia fetida* is commonly known as the compost worm, wiggler or redworm. They are brown, red or purple in colour and distributed throughout the country due to their migratory behavior (Talashilkar and Dosani, 2008). *Eisenia fetida* can tolerate temperature up to 30-35°C (Garg *et al.,* 2005). Cattle dungs are suitable culture medium for *Eisenia fetida* (Aalok *et al.,* 2008). They are able to consume food more than that of their body weight each day (Elvira *et al.,* 1998. *Eisenia fetida* have short life span; yet very active with high regenerative capacity.

The cast production efficiency of *Eisenia fetida* ranges from 8-12 mg/worm per day. Mature adult may have body weight from 0.7-1.5 g (Srivastava and Singh, 2004). The compost worm need five basic things: a hospitable living environment (bedding), a food source, adequate moisture (Greater than 50 per cent water content by weight), an adequate aeration and production (Card *et al.,* 2004). They grow faster and under the optimum condition of temperature, humidity and food quantity they attained reproduction capability within 30-40 days (Gupta, 2005; Bhartiya

**Figure 3**: *Eisenia fetida.*

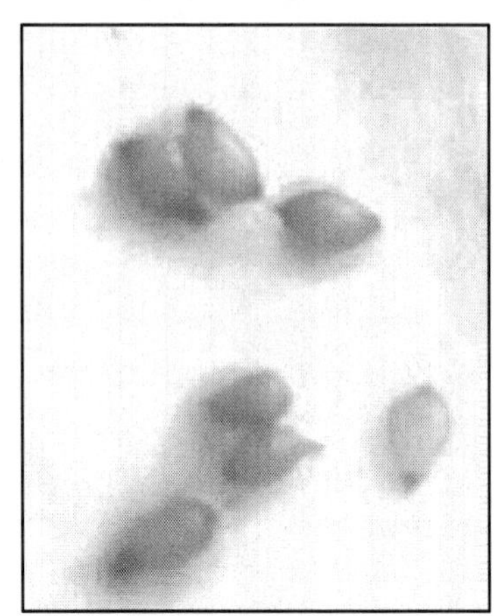

**Figure 4**. Cocoon of *Eisenia fetida.*

*et al.,* 2011). Mature adult worm produces 2-3 cocoons per weeks (Lowe and Butt, 2002). The egg hatched within 3 weeks and each cocoon produces 2-20 baby worms (Ghatnekar *et al.,* 1998). An adult *Eisenia fetida* produces approximately 250 worms within 6 month. The life span of *Eisenia fetida* is 70 days (Gupta, 2005).

The dung material strongly influences the biology of *Eisenia fetida* which are used in vermiculture operation and manure management of dairy, horse and rabbit (Card *et al.,* 2004). Kaviraj and Sharma (2003) reported the *Eisenia fetida* to be superior in over Lampitto mauritii for conversion of waste in to vermicomposts. Garg *et al.* (2006 b) and Suthar (2007 b) observed that the *Eisenia fetida* have a great potential to change the different live stock excreta in vermicomposts. Earthworm show better growth and fecundity in pre-composted cattle solid wastes (Gunadi *et al.,* 2002). Biomass gain and cocoon production by *Eisenia fetida* was more in cow, buffalo dung than goat dung (Loh *et al.,* 2005). Gunadi and Edwards (2003) observed the fecundity and mortality of *Eisenia fetida* in different manure for more than one year and found that the worm growth faster in pig wastes than cattle solid wastes. Addition of cow dung with solid textile mill sludge is suitable for the survival of *Eisenia fetida* (Kaushik *et al.,* 2003, 2004). It also has great impact on nitrogen transformation (Atiyeh *et al.,* 2000).

Large amount of garbage generated by the human population in the form of solid, liquid and organic wastes can be used for vermi-treatment and converted into utilizable end products. The application of earthworm is very effective method to control the organic wastes generated from household to industrial unit (Trivedi and Kumar, 1998; Taylor *et al.*, 2003). Earthworm as a potential source for biodegradation of organic waste such as pig, horse manure, rabbit droppings, crop residue and city garbage (Ghatnekar *et al.*, 1998; Abbassi and Ramasammy, 2001).

Earthworms concentrated the pesticides and heavy metals in their tissues. Sodium (Na) was concentrated by *Eisenia fetida* (Garg *et al.*, 2005; Gupta *et al.*, 2005). Mature adults of Lumbricus rubellus have higher level of cadmium than immature adults. The bioaccumulation of toxicants in the tissues of earthworms varies and depending on the soil properties like-pH and calcium concentration (Govindan, 1998; Bhartiya and Singh, 2012a).

Earthworm feeds on waste biosolid per day up to twice their bodyweight and makes possible to convert the biological wastes into vermicompost (Haimi and Hutha, 1988). The enzyme produced by earthworms and micro-organisms play an important role in soil fertility. Worm cast enhanced the microbial activity, resulted the increase of enzyme activity, micro and macro-nutrients in the soil (Suthar, 2006 a; Suthar, 2008 b). It is also reported that microbial group in the vermicomposting produced an intercellular dehydrogenase and phosphates enzyme accelerate the oxidative phosphorylation process (Garcia *et al.*, 1999; Masciandro *et al.*, 2000; Madjen *et al.*, 2001).

*Eisenia fetida* degrade wastes in to vermicompost and produced vermitien, useful for fish bait. Earthworms increase the nitrate production by stimulating bacterial activity, mucus production, and dead tissues decomposition (Aira *et al.*, 2002). Feeding on aerobic sewage sludge and domestic animal manure, earthworm increased the rate of decomposition (Chaudhary and Bhattacharjee, 2002; Kaushik *et al.*, 2003). Mitchell (1997) observed that during vermicomposting, earthworm decrease the anaerobic process and increase the aerobic condition, therefore decline in methane and volatile sulphur compounds.

Garg *et al.* (2005) reported that cow, horse, goat and sheep dung supported the growth and reproduction of *Eisenia fetida*, hence it can be used as feed materials in large scale vermicomposting. Purohit, (2003) and Garg *et al.* (2006 a) studied that the combination of water hyacinth and cow dung retarded the growth and fecundity of earthworm *Eisenia fetida* and also affect the nutritional quality of vermicompost. In the large scale vermicomposting, the use of solid textile mill sludge as raw material with inoculation of *Eisenia fetida* can help to convert of these wastes in to value added product (Kaushik *et al.*, 2003; Garg *et al.*, 2005).

# 11

# Therapeutic Uses of Earthworms

Stephenson (1930) has reported that earthworm ashes used as tooth powder, as stimulant for hairs growth in head. It is used in the treatment of piles, fever, small pox, jaundice and removal of stones in bladder (Ranganathan, 2006). Barley (1961) reported earthworms are used as antipyretic. Earthworm and their extract have anti-oxidant activity, cure impotency, rheumatism (Weisbach, 1962), promote lactation, and dilate the bronchi (Reynolds and Reynolds, 1972).

Prakash (2006) has reported that the administration of earthworm paste of *Lampito mauritii* had restored the gastero-intestinal damage by reducing the gastric acid secretion, acidity and enhancing the pH. It had also increased the activity of anti-oxidative enzymes prevent the damage of mucous membrane in the stomach of albino rat. The paste and its extract of earthworm *Eisenia fetida* prevent the oxidative damage because the earthworm tissue have significant amount of anti-oxidant such as glutathione and glutathione peroxides. The paste of *Lempitto mouritii* was found to enhance the liver antioxidant such as GSH (reduced glutathione),

GPx (Glutathione peroxidase) and CAT (catalase) and decrease the lipid per oxidation in albino rat (Balamurugan, 2006).

Various extract of earthworm posses potency against some pathogenic and non pathogenic bacteria such as *Eischerichia coli, Streptococcus pyogenes, Pseudomonas aeruginosa* and *Salmonella entiertidi* (Viallier *et al.*, 1985). Popovic *et al.* (2005) have reported the anti-bacterial activity of *Eisenia fetida* against Streptococcus pyogenes, Pseudomonas aeruginosa etc. Earthworm paste and its extract have anti-inflammatory properties in both acute and chronic phase (Balamurugan, 2006; Yegnanarayan *et al.*, 1998). Nagasawa *et al.* (1991) reported that the skin extract lumbricine from *Lumbricus terrstris* inhibit the growth of mammary tumors in SHN mice. Yegnanarayan *et al.* (1987) found that ethanol and petroleum extract of earthworm *Lempito mauritii* exhibit anti-inflammatory activity in albino rat. Herzenjak *et al.* (1992) extracted a biologically active glycoprotein G-90 from whole earthworm tissue (*Eisenia fetida, Lumbricuss rubellus*) homogenate and found that it slow down the tumor growth mouse. Earthworm's coelomic fluid and its tissue extract exhibits a strong anti microbial activity. Popovic *et al.* (2001) isolated the proteolytic enzyme *Eisenia fetida* which caused lysis of clots originated from venous blood of dog with cardiopathies and with malignant tumors. Herzenjak *et al.* (1998) have reported the anti-coagulant properties of earthworms. They isolated serine protease as anti coagulant.

It is clear that in spite of growth enhancer properties they also affect the early germination period, growth, early flowering period and productivity of plants. The chemical analysis provides the knowledge of chemical composition of each initial feed mixture, final vermicomposts and mixture of soil with different vermicomposts. It will be more beneficial to the particular type of vermicomposts for particular nutrient deficient soil. Vermicomposts is easily producible, easily biodegradable, less expensive, and no hazards on soil, human health and environment. It significantly decreases the germination period, flowering period and increase the growth and productivity as well as safe for soil, environment and human health. It use will be ecologically safe and culturally more acceptable among farmers and can improve the socio-economic conditions of farmers in rural areas.

The vermiwash is also a potent source for plant growth and suppression of various microbial diseases. Although a lot of work on vermiwash has been reported, yet, there is no study on the vermiwash produced from different wastes products singly and their binary combinations on the plant growth and crop yield. In the present study attention has been focused on the preparation of different type of vermiwash by worms grown in different animal wastes such as cow, buffalo, goat, sheep, horse and agro wastes singly and in binary combinations. Effect of these vermiwash on growth, flowering and productivity of certain crops will be studied. Chemical analysis of vermiwash of different combinations of these wastes will be performed to correlate their effect on plant growth. It is hoped that data emerging from this study will be an efficient biotechnological tool (liquid biofertilizers) obtained from different wastes for enhancement of plant growth and yield.

# 12

# Methods of Vermicomposting, Vermiwash and Biopesticides

## 1. Collection of Wastes

### (a) Animal Wastes

The fresh animal dung *viz*-cow, buffalo, horse, sheep and goat dung were collected from different animal farms of the Gorakhpur district.

### (b) Agro/Kitchen Wastes

The organic wastes (agro and kitchen) were collected from the garbage and agricultural field of rural and urban areas of Gorakhpur, U.P. India.

## 2. Culture of Earthworms

Healthy adult epigeic earthworms, reddish brown in colour, commonly known as red wiggeler worm, *Eisenia fetida* are cultured in vermiculture research centre, Department of Zoology, D.D.U. Gorakhpur University, Gorakhpur, U.P. India. *Eisenia fetida* procured from this centre. They donot process the soils but are efficient in composting of organic wastes. They enhance the area of organic manure production through biodegradation or mineralization and nutrient mobilization. When maintained in captivity under semi-natural conditions, they remain active throughout the year. *Eisenia fetida* dominates in soils of pH 7 to 8. The temperature tolerance capasity of *Eisenia fetida* ranged from 20-30 (C and aeration, moisture up to 40 per cent to 60 per cent were maintained for proper growth and survival of earthworms.

## 3. Experimental Setup for Vermicomposting

The vermicomposting were conducted on cemented earth surface. There are 35 vermibed formed by different combinations of animal, agro/ kitchen wastes in 1:1 ratio the size of each vermibed is 3m x 1m x 9cm. After formation of vermibed moist 2kg of cultured *Eisenia fetida* were inoculated in each bed. The beds were covered with jute pockets and moisten the bed daily up to 40 to 50 days for maintaining the moisture content. After one week interval, mixture of bed was manually turned up to 3 weeks. After 45 to 50 days granular tea like vermicompost appear on the upper surface of beds.

### 4. Extraction of Vermiwash

Vermiwash extracted from vermiwash collecting device by the method of Ismail, 1997). The apparatus was made from plastic or metals drum having capacity of 2 liter and a tap at the bottom of the drum filled with

**Figure 4**: Vermiwash Collecting Device.

broken breaks, about 3cm thickened which is followed by sand layer of 2-3 cm thickness lastly with filled with vermicompost with heavy population of earthworms, simultaneously fresh water was added in to the drum and a container kept bellow the tap of drum. The watery extract of vermicomposts, vermiwash drainage out off drum. The colour of vermiwash ranges from yellowish to black. After 1 to 2 days the process of extraction has been completed.

## 5. Collection and Prepration of Biopesticides

Biopesticides prepared from the plants of neem, annona and garlic through extraction method in the form of oil, powder and extract by the method of Singh and Singh, 1995; and Singh and Singh, 2001 respectively.

### i. Prepration of Neem Extract

Fresh leaves (250 mg) of neem (*Azadirachta indica*), were crushed in 250 ml water. Extract was filtered out and fridged. According to need, extract with vermiwash (w/v) in different ratios, sprayed on crops.

### ii. Prepration of Garlic Extract

Bulbs of garlic (*Allium sativum*), were crushed and extract was filtered out and fridged. After diluting with water, applied on different crops, with addition of vermiwash.

### iii. Prepration of Annona (*Annona squemosa*) Extract

Fresh leaves (250 mg) of annona (*Annona squemosa*) were crushed in 250 ml of water. Extract was filtered out and fridged. During application, mixed with different vermiwash combinations and sprayed on different crops.

### iv. Multineem (Neem Oil)

Azadiractin 0.03 per cent, 90.57 per cent neem oil, 5 per cent hydroxyl, 0.50 per cent epichlorohydrine and 3.0 per cent aromax; is purched from Agricare Pvt. Ltd. Karnatka, India.

Preparation of vermiwash and biopesticide mixture in different ratio- Mixture of vermiwash and biopesticides applied on crops in 10:1, 20:1 and 10:3 for growth observation and in the same manner, 10:1, 10:2, 10:3 ratio of mixture applied for pest infestation control.

## 6. Chemical Analysis of Vermicompost and Vermiwash

The following methods were used for the evaluation of different type of chemical content.

### i. Physico-chemical Parameters Include

- ☆ Total Organic Carbon (TOC)
- ☆ Total Kjeldahl Nitrogen (TKN)
- ☆ C: N ratio
- ☆ Total Phosphorus (TAP)
- ☆ Total Potassium (K)
- ☆ Total Calcium (Ca)
- ☆ pH

### ii. Electrical Conductivity (EC)

Samples extract preparation and chemical analysis was done by the methods of Trivedi and Goel (1984). The details of methods are given as follows:

### iii. Total Kjeldahl Nitrogen (TKN)

Total Nitrogen was determined by microkjeldahl method (Bremner and Mulvaney, 1982).

**Reagents**

Digestion mixture – It was prepared by mixing $K_2SO_4$, $CuSO_4$ and $SeO_2$ in the ratio of 10: 4: 1.

Boric acid indicator solution: Twenty gram boric acid was dissolved in about 700 ml of hot distilled water. Solution in then cooled and transferred to one liter volumetric flask containing 20 ml of mixed indicator solution (prepared by dissolving 100 mg bromocresol green and 50 mg of methyl red in 100ml of ethanol). After mixing the contents of flask, the solution was diluted to one liter with water and mixed thoroughly.

Sulphuric acid: AR grade concentrated sulphuric acid.

HCl: N/50 HCl was prepared for titration.

**Procedure**

500 mg of dry material was taken in digestion tube. To each tube, 1 gm of digestion mixture and 10 ml of sulphuric acid was added. It was

then digested on Kjeldahl digestor till bluish green colour appeared in tube. Then 10 ml of boric acid indicator solution was taken in 100 ml Erlenmeyer flask that was marked to indicate a volume of 50ml and flask was placed under the condensor of steam distillation apparatus. The digested mixture was taken in distillation flask and flask was attached to steam distillation apparatus. Excess of 90 per cent NaOH was added to tube until green color turned black. Distillation was started by closing stop cork on steam by pass tube of distillation apparatus. When the distillate reached 50ml mark of receiver flask, the distillation was stopped by opening stop cork on steam by pass tube and end of condensor was rinsed. The total N in distillate was determined by titration with N/50 HCl. The color change at end point was from green to permanent faint pink.

**Calculation**

$$\text{N (per cent)} = \frac{V \times 0.00014 \times D \times 10}{W \times A}$$

*where,*

V: Volume of N/50 HCl used.

D: Dilution factor (Volume made in volumetric flask)

W: Weight (g) of sample.

A: Aliquot taken

## iv. Total Organic Carbon (TOC)

Organic carbon was determined by dry combustion (Nelson and Sommers, 1982). Weighed 500 mg of dried ground (<2mm) sample into preweighed china crucibles. The sample was ignited in a muffle furnace at 6000C for 1½ h. The furnace was allowed to cool and the ash produced was weighed. Organic C was calculated from the following relationship:

Organic C (per cent) = (100 – ash per cent)/1.724

## v. Carbon: Nitrogen (C:N) Ratio

$$\text{C/N ratio} = \frac{\text{Total concentration of carbon}}{\text{Total concentration of nitrogen}}$$

## vi. Total Phosphorus (TP)

This was determined by spectrophotometric method.

### Reagents

0.002N $H_2SO_4$ solution

Ammonium molybdate solution : (a) 2.5g of ammonium molybdate was dissolved in 15 ml of distilled water (b) 28 ml of concentration $H_2SO_4$ was diluted with distilled water to make it 40 ml. Solution (a) and (b) were mixed and final volume was made 100 ml with distilled water.

Stannous chloride solution: 2.5 gm of $SnCl_2$ was dissolved in 10 ml of HCl and volume was made 100 ml.

### Procedure

0.5gm sample was taken in 250 ml conical flask and to it added 100 ml of 0.002N $H_2SO_4$. Solution was shaken for half an hour and then added 2 ml of ammonium molybdate solution. Took 50 ml of filtered solution. Then 5 drops of $SnCl_2$ were added. A blue colour appeared and reading was taken at 690 nm on a spectrophotometer using a blank with same amount of reagents. Reading was taken after 5 minutes but before 12 minutes of the addition of last reagent. The concentration was found out with help of standard curve.

### Calculation

Avalable phosphorus (per cent) = P x V/W x 100

*where,*

P: Phosphate content in extract (mg/l).

V: Total volume of extract (ml) prepared.

W: Weight of air dried sample (g).

## vii. Potassium (K)

This was determined by flame photometric method.

### Reagent

Ammonium acetate solution.

### Procedure

0.5 gm of sample was weighed and to it added 100 ml of ammonium acetate solution and kept for overnight. Now solution was filtered and during filtration first few ml of filtrate may be discarded. Now potassium content in extract was determined by using potassium filters after necessary settings and calibration of the instrument. Readings for sample were noted. Potassium content in sample was determined with the help of standard curve.

**Calculation** Available Potassium $= \frac{K \times V}{10000 \times S}$

*where,*

V: Total volume of sample extract prepared.

S: Weight of sample taken

### viii. Calcium (Ca)

Optical filter was used, in the flame photometric determination of calcium (Ca) in the aliquot of the neutral N $NH_4OAc$ extract. Set the zero point of galvanometer with distilled water and 100 marks with 20 ppm of calcium solution. Make the calibration curve of Ca, using the flame photometer readings of the standards on the ordinate and the concentrations (ppm) on the abscissa. Read the sample, after necessary dilution with distilled water, and after verifying from time to time with 20 ppm Ca solution to set the 100 mark.

### ix. pH

A sample suspension was made by taking 10 g of air dried sample with 50 ml distilled water in a flask to prepare a suspension (1:5 w/v). The suspension was kept over a shaker for 30 minutes. The shaking period was not allowed to exceed 30 minutes otherwise various biological processes may start in suspension which may change actual pH of suspension. The pH of suspension was estimated using pH meter. The pH meter was calibrated with buffer solution of pH 4, 7 and 9.2 prior to its use.

### x. Electrical Conductivity

A sample suspension was made and used. To prepare soil suspension, 10 gm of air dried sample was added to 50 ml of distilled water in a flask and it was shaken well on a shaker for half an hour. The conductivity meter was adjusted to known conductivity with standard KCl solution (0.1N) by cell constant. The conductivity is measured by dipping electrode into the solution.

## 7. Experimental Setup

Measurement of seed germination period, growth, flowering period and productivity of crops were performed in the experimental field. The different varieties of crops were showed according to their season. viz-rabi, kharif and zayad. They were sowed directly in the cultivated soil. In the cultivated field, each square having the size of 1m2 (1m x 1m) area seed of different crops were showed in each squire in equal amount.

## 8. Measurement of Germination Period, Growth, Flowering Period and Productivity

Germination period were determined by appearance of new seedling on the upper surface of soil. Growth was observed at 20, 35 and 50 days by auxanometer. Flowering period (days) was observed when flower appear in adults plants. After harvesting of each crops, productivity were calculated in the kg/meter$^2$.

## 9. Statistical Analysis

The data given as the mean ± SE of six replicates. Two way analysis of variance (ANOVA) were applied in between initial feed mixture and final vermicomposts of different combinations of animal, agro and kitchen wastes for chemical analysis as well as the significant growth of various crops between different vermicomposts. Student 't' test were applied in between control and mixture of soil with different vermicompost for germination, flowering and productivity (Sokal and Rohlf, 1973).

# DIFFERENT VEGETABLE CROPS

**Figure 5**. Potato (*Solanum tuberosum*).

**Figure 6**. Tomato (*Lycopersicon esculentum*).

**Figure 7**. Cauliflower (*Brassica oleracea* var. *botrytis*).

**Figure 8**. Bean (*Lablab purpureus purpureus*).

**Figure 9**. Brinjal (*Solanum melongena*).

**Figure 10**. Okra (*Abelmoschus esculentus*).

# References


Aalok, A., Tripathi, A.K. and Soni, P. (2008). Vermicomposting: A better option for organic solid waste management. J. Hum. Ecol., 24(1), 54-64.

Abbasi, S.A. and Ramasamy, E.V. (2001). Solid Wastes Management with earthworms. Discovery Publishing House, New Delhi, India, pp. 78.

Adriano, D.C. (2001). Trace element in terrestrial Environments; Biochemistry, Bioavialibity and Risk of metals, Springer- Verlage, NewYark.

Agrawal, S.B., Singh, A. and Diwedi, G. (2003). Effect of vermicompost, Farm yard manure and chemical fertilizers on growth and yield of wheat (*Triticum aestivum* L.) Variety HD-2643. Plant Arch., 3 (1), 9-14.

Aira, M., Monroy, F., Dominguez, J. and Mato, S. (2002). How earthworm density affects microbial biomass and activity in pig manure. Eur. J. Soil Biol., 38, 7-10.

Albanell, E., Plaixats, J. and Cabrero, T. (1988). Chemical changes during vermicomposting (*Eisenia foetida*) of sheep manure mixed with cotton industrial wastes. Biol. Fertil. Soil, 6, 266-269.

AL-Dahmani, J.H., Abbasi, P.A., Miller, S.A. and Hoitink, H.A.J. (2003). Suppression of bacterial spot of tomato with foliar sprays of compost extracts under green house and field conditions. Plant Disease, 87, 913-919.

Ansari, A.A. (2007). Reclamation of sodic soils through vermitechnology. Green world foundation, Dhaka, Bangladesh, Journal of Soil and Nature, 1 (1), 27-31.

Ansari, A.A. (2008). Soil profile studies during bioremediation of sodic soil. Through the application of organic amendment (vermiwash, tillage green manure, mulch, earthworm and vermicompost), World J. of Agriculture Sciences, 4 (5), 550-553.

Appelholf, M. (2003). "Notable Bits". In Wormezin, Vol 2(5).

Appelholf, M., Edwards, C.A. and Neuhauser, E. F. (1998). Domestics vermicomposting systems. Earthworm Waste Environ. Manage., pp. 157-161.

Arancon, N.M., Edwards, C.A., Bierman, P., Metzger, J.D., and Welch, C. (2005). Effects of vermicomposts produced from cattle manure, food waste and paper waste on growth yield of peppers in the field. Pedobiologia, 49, 297-306.

Arancon, N.Q., Edwards, C.A., Bierman, P., Metzger, J.D., Lee, S. and Welch, C. (2003). Effects of vermicomposts on growth and marketable fruits of field-grown tomatoes, peppers and strawberries. Pedibiologia, 47, 731-735.

Arancon, N.Q., Edwards, C.A., Bierman, P., Welch, C., and Metzger, T.D. (2004). Influences of vermicomposts on field strawberries. Effect on growth and yields. Biores. Technol., 93, 145-153.

Arguello, J. A., Ledesma, A., Selva, B., Nunez, Carlos, Rodriguez, H., Mariadel, C. and Diaz G. F. (2006). Vermicomposts effect on bulbing dynamics, Nonstructural corbohyadrates Content, Yield and quality of *Rosado paraguayo* (Garlic bulb). Hort. Science, 41(3), 589-592.

Arunachalam, S. and Palanichamy, S. (1984). Earthworms as a feed for the catfish *Mystus vittatus*. A paper presented to national seminar on organic waste utilization and vermicomposting. School of Life Science Sambalpur University Jyoti Vihar, Orissa, pp. 4.

Ashok, K.C. (1994). State of art report on vermicomposting in India.Council for advancement of People Action and Rural Technology (CPART), New Delhi, pp.60.

Atharasopoulous, N. (1993). Use of earthworm biotechnology for the management of aerobically stabilized effluents of dried vine industry. Biotechnol. Lett., 15 (12), 126-128.

Atiyeh, M.R., Dominguez, J., Subler, S., and Edwards, C.A. (2000). Changes in biochemical properties of cow manure during processing by earthworms and the effects on the seedling growth. Pedobiologia, 44, 709-724.

Atiyeh, R.M., Arancon, N.Q., Edwards, C.A. and Metzger, J.D. (2002). The influence of humic acid derived from earthworms processed organic wastes on the plant growth. Biores. Technol., 84, 7-14.

Atiyeh, R.M., Edwards, C.A., Sublar, S. and Metzger, T. (2001). Pig manure vermicompost as a component of a horticultural bedding plant medium. Effects on physiochemical properties and plant growth. Biores. Technol., 78, 11-20.

Atiyeh, R.M., Subler, S., Edwards, C.A. and Metzger, J.D. (1999). Growth of tomato plants in horticulture potting media amended with vermicompost. Pedobiologia, 43, 1-5.

Atlavinyte, O., Baydonaviciena, Z. and Bada-Vicience, I. (1968). Earthworm for soil health and pollution control. J. Sci. Ind. Res., 42, 575-583.

Azarmi, R., Giglou, M.T. and Taleshmikail, R.D. (2008). Influence of vermicompost on soil chemical and physical properties of tomato (*Lycopercicum esculentum*) field. African J. of Biotech., 7(14), 2397-2401.

Balaji, S. (1994). Studies on vermicomposting and its influence on plant growth. M.Phil. Thesis, Ann. University, Madras, India.

Balam, (2000). Studies on biopesticidal activity of vermiwash in control of some foliar pathogens. M.Sc. (Agri) Thesis submitted to Dr. B.S.K.K.V. Dapoli.

Balamurugan, M. (2006). Effect of earthworm paste *Lempito maurtii*, (Kinberg) on the anti inflammatory, antioxidative, haematological and serum biochemical indices of rat (*Rattus norvegicus*) M. Phil., Thesis, Annamalai University.

Bansal, S. and Kapoor, K. K. (2000). Vermicomposting of crop residues and cattle dung with *Eisenia foetida*. Biores. Technol., 73, 95-98.

Barley, K.P. (1961). The abundance of earthworm in agricultural land and their possible significance in agriculture. Adv. Agron., 13, 249-268.

Bawa, J.N. (1995). Studies on the effect of different organic manures, chemical fertilizers and growth promoter on growth, yield and quality of rabi hot weather groundnut (*Arachis hypogea* L.) under leteritic soils of konkan region. M.Sc. (Agri.) Thesis, K.K.K. Dapoli.

Beetz, A. (1999). Worms for composting (vermicomposting). Livestock technical note, ATTRA, Fayetteville, AR.

Benitez, E., Nogales, R., Elvira, C., Masciandaro, G. and Ceccanti, B. (1999). Enzyme activities as indicators of stabilization of sewage sludges composting with *Eisenia foetida*. Biores. Technal., 67, 297-303.

Benitez, E., Nogales, R., Masciandro, G. and Ceccanti, B. (2000). Isolation by isoelectric focusing of humic–urease complexes from earthworm (*Eisenia foetida*) processed sewage sludges. Biol. Fert. Soils, 31, 489 – 493.

Benitz, E., Sianz, H. and Nogales, R. (2005). Hydrolytic enzyme activities of extracted humic substances during the vermicomposting lignocelulosic olives waste. Biores. Technol., 96, 785-790.

Bhartiya D. K. and Singh K. (2011). Accumulation of Heavy Metals by *Eisenia foetida* from Different animal dung and Kitchen wastes during Vermicomposting International Journal of Life Science and Technology. 2011, 4 (7): 47-52.

Bhartiya D. K. and Singh K. (2012a) Heavy Metals Accumulation from Municipal solid wastes with Different animal dung through Vermicomposting by Earthworm *Eisenia fetida*, World Applied Sciences Journal 17 (1): 133-139.

Bhartiya D. K. and Singh K. (2012b). Heavy Metals Accumulation from Municipal solid wastes with Different animal dung through Vermicomposting by Earthworm Eisenia fetida. American-Eurasian J. Agric. and Environ. Sci., 12 (9): 1215-1222.

Bhartiya D. K., Nath G and Singh K. (2011) Vermicomposting; A tool for management of different wastes and self-employment for youth of B.P.L. and weaker sections. Proceeding of National Seminar on 'Challenges for Biosciences in 21st Century. Department of Zoology, S. P. P. G. College Shohratgarh, Siddharthnagar, pp 71-73.

Bhatnagar, R.K. and Palta, P.K. (1998). Vermiculture and vermicomposting. Kalyani Publication, pp. 101.

Bhattacharya, P. (2004). Organic food production in India. Agrobios (India), Jodhpur.

Bhattacharya, S.S. and Chattopadhyay, G.N. (2004). Transformation of nitrogen during vermicomposting of fly ash. Waste Management Resource, 22, 488 - 491.

Bhawalkar, U.S. (1990). Bioconversion of wastes in to resources 4th International Symposium on earthworm ecology. June 10-14, 1990 Avignon, France.

Bhole, R.J. (1992). Vermiculture biotechnology basic scope for application and development. Proc. National Seminar on Organic Farming held at college of Agril., Pune, Jan., 28-29.

Bisoyi, R.N. (2003). Potentialities of organic forming in India RBDC, Bangalore.

Board, N.I.I.R. (2008). The complete technology book on vermiculture and vermicompost. National Institute of Industrial Research, New Delhi, India, pp. 1.

Bouche, M.B. (1977). In soil organisms as components of ecosystems. (U. Lohm, and T. Person, eds.) Ecol. Bull., 25, 122-132.

Bouche, M.B. (1987). Emergance and development of vermiculture and vermicomposting from a hobby to an industry, from marketing to biotechnology, from irrational to credible practices. (A.M. Bonvicini, Pagliai and P. Omodeo eds.) on earthworm. Slected Symposia and Monograph U.Z.I., 2 Muchhi, Modena, pp. 519-531.

Brady, N.C. and Weil, R.R. (2002). The nature and property of soil. 13th Ed. Prentice Hall, Upper Saddle River, N.J., pp. 960.

Bremner, J.M. and Mulvaney, R.G. (1982). Nitrogen Total in Method of Soil Analysis (A.L. Page, R.H. Millar and D.R. Keeney, eds.), American Society of agronomy, Madison, pp. 575-624.

Buckerfield, J.C. and Webster, K.A. (1998). Worm-worked waste boosts grape yields: Prospects for vermicompost use in vineyards. Australian and New Zealand Wine Industry Journal, 13, 73-76.

Buckerfield, J.C., Flavel, T.C., Lee, K.E. and Webster, K.A. (1999). Vermicompost in solid and liquid form as plant growth promoter. Pedobiologia, 43, 753-759.

Buckerfield, J.C., Flavel, T.C., Lee, K.E. and Webster, K.A. (1999). Vermicompost in solid and liquid form as plant growth promoter. Pedobiologia, 43, 753-759.

Bulluck, L.R. and Ristaino, J.B. (2002). Effect of synthetic and organic soil fertility amendments on shouthern blight, soil microbial communities, and yield of processing tomatoes. Phytopathology, 92, 181-189.

Butt, K.R. (1993). Utilization of solid paper mill sludge and spent brewery yeast as a feed for soil dwelling earthworms. Biores. Technol., 44, 105-107.

Butt, K.R., Frederickson, J. and Morris, R.M. (1995). An earthworm cultivation and soil inoculation technique for land restoration. Ecol. Eng., 4 (1), 1–9.

Butt, K.R., Fredrickson, J. and Morris, R.M. (1992). The intensive production of *Lumbricus terrestris* L. for soil amelioration. Soil Biology and Biochemistry, 24, 1321-1325.

Canellas, L.P., Olivares, F.L., Olorolova, A.L. and Facanda, A.R. (2002). "Humic acids isolated from earthworm compost enhance root elongation, areal root emergence, and plasma membrane $H^+$ $ATP_{ase}$ activity in maize root". Plant Physiology, 130, 1951-1957.

Card, A.B., Anderson, J.V. and Davis, J.G. (2004). Vermicomposting horse manure. Colorado State University Cooperative Extension no., 1.224.

Chan, C., Belanger, R.R., Benhamou, N. and Poulitz, T.C. (2000). Defence enzyme induced in cucumber roots by treatments with plant growth-promoting rhizobacteria (PGPR) and *Pythium aphanidermatum*. Physiological Molecular Plant Pathology, 56,13-23.

Chan, P.L.S. and Griffiths, D.A. (1988). The vermicomposting of pretreated pig manure. Biol. Wastes, 24, 57-69.

Chaoui, H., Edwards, C.A. and Brickner, M. (2002). Suppression of plant disease, pythium (damping off), rhizoctonia (root rot) and verticillum (wilt) by vermicompost proceedings of brighton crop protection conference-pest and diseases, Vol. II (8B), 711-716.

Chaoui, I., Zibiliske, M. and Ohno, T. (2003). Effect of earthworms cast and compost on soil microbial activity and plant nutrient availability. Soil Boil. Biochem., 35, 295-302.

Chaoui, I., Zibiliske, M. and Ohno, T. (2003). Effect of earthworms cast and compost on soil microbial activity and plant nutrient availability. Soil Boil. Biochem., 35, 295-302.

Chaudhari, P.S. (2005). Vermiculture and vermicomposting as biotechnology for conservation of organic waste in to animal protect and organic fertilizer. Asian J. Micro., Biotech. and Environ. Science, 7, 359-370.

Chaudhari, P.S. and Bhattacharjee, G. (2002). Capacity of various experimental diets to support biomass and reproduction of Perionyx *excavatus*. Biores. Technol., 82, 147-150.

Contreras-Ramos, S.M., Escamilla-Silva, E.M. and Dendooven. (2004). Vermicomposting of biolisolids with cow manure and oat straw. Biol. Fertile Soil, 41, 190-198.

Crescent, T. (2003). Vermi composting, development alternatives (DA) sustainable livelihoods (http://www. dainet. org/livelihoods diffault. htm)

Crossley, D.A. Reichele, D.E. and Edwards, C.A. (1971). In take and turn over of radio active Cesium by earthworms (Lumbricidae). Pedobioloigia, 11, 71-76.

Darwin, C. (1881). The formation of vegetable mould through the action of earthworms with observation of their habits (Murray), Landon, pp. 326.

Davis, J.G., Lverson, K.V. and Vigi, M.F. (2002). Nutrient variability in manure: Implications for sampling and regional database creation. Journal of soil and water conservation, 57 (6), 473-478.

Desai, S.S. (2003). Effects of city compost, sewage sludge, and vermiwash on the flower, yield, nutrient uptake and keeping quality of china-aster (*Callistephus chinensis*). M.Sc. (Agri.) Thesis submitted to Dr. B.S. Konkan Krishi Vidyapeeth, Dapoli

Desai, V.R., Sabanel, R.N. and Rundal, P.V. (1999). Integrated nitrogen management in wheat-coriander cropping system. Journal of Maharasthra Agricultural Universities, 24 (12), 273-275.

Devi, D. and Agarwal, S.K. (1998). Performance of sunflower hybrids as influenced by organic manure and fertilizer. Journal of Oilseeds Research, 15(2), 272-279.

Dhawan, K., Malhotra, S., Dahiya, B.S. and Singh, D. (1991). Seed protein fractions and amino acid composition in gram (*Cicer arientinum*). Plant Food for Human Nutrition, 41, 225-232.

Eastman, B.R., Kane, P.N., Edwards, C.A., Trytek, L., Gunadi, B., Stermer, A.L. and Mobley, J.R. (2001). The effectiveness of vermiculture in human pathogen reduction for USEPA biosolids stabilization. Compost science and utilization, 9, 38-49.

Edwards, C.A. (1998). The use of earthworms in the breakdown and management of organic wastes. (In: C.A. Edwards eds. Earthworm Ecology). CRC Press, Boca Raton, FL, pp. 327-354.

Edwards, C.A. and Baxter, J.E. (1992). The use of earthworm in environmental management. Soil Biology and Biochemistry, 24 (12), 1683-1689.

Edwards, C.A. and Bohlen, P.J. (1996). Biology and Ecology of Earthworm. (3rd Eds.), Chapman and Hall, London, pp. 426.

Edwards, C.A. and Burrows, I. (1988). The potential of earthworm compost as plant growth media. (In: C.A. Edwards and E.F. Neuhauser eds. Earthworms and Waste Environmental Management) SPB Academic Publishing, The Haque, Netherlands, pp. 211-220.

Edwards, C.A. and Lofty, J.R. (1972). Biology of earthworm Chapman and Hall London pp. 333.

Edwards, C.A. and N. Arnacon. (2004). Vermicomposts suppress plant pest and disease attacks. In REDNOVA NEWS: http://www.rednova.com/display/id=55938.

Edwards, C.A., Bohlen, P.J., Linden, D.R and Subler, S. (1995). Earthworms in agro ecosystems. (In: Earthworm Ecology and Biogeography in North America. P.F. Hendrix,eds.), Lewis Publisher, Boca Raton, FL, pp, 185-213.

Edwards, C.A., Dominiguez, J. and Arancon, N.Q. (2004). The influence of vermicompost on plant growth and pest incidence. (In: S.H. Shakir and W.Z.A. Mikhail, eds.). In Soil Zoology for Sustainable Development in the 21st century), Self Publisher, Cairo, Egypt. pp. 397-420.

Eghball, B. and Gilley, J.E. (1999). Phosphorus and nitrogen in runoff following beef cattle manure or compost application. Journal of Environmental Quality, 28, 1201-1210.

Elvira, C., Goicoechea, M., Sampedro, L., Mato, S. and Nogales, R. (1996). Bioconversion of solid paper-pulp mill sludge by earthworms. Biores. Technol., 75, 173-177.

Elvira, C., Sampedro, L., Benitez, E. and Nogales, R. (1998). Vermicomposting of sludges from paper mill and dairy industries with *Eisenia andrei*: A pilot scale study. Biores. Technol., 63, 205–211.

Elvira, C., Sampedro, L., Dominguez, J. and Mato, S. (1997). Vermicomposting of waste water sludge from paper-pulp industry with nitrogen rich materials. Soil Biol. Biochem., 29, 759-762.

Fang, M., Wong, J.W.C., Ma, K.K. and Wong, M.H. (1999). Co- composting of sewage sludge and coal fly ash: nutrient transformations. Bioresource Technology, 67, 19-24.

Fokkema, N.J. (1993). Opportunities and problems of control of foliar pathogens with microorganisms. Pesticide Science, 37, 411-416.

Frederickson, J., Butt, K.R., Morris, R.M. and Danial, C. (1997). Combining vermiculture with traditional green waste composting systems. Soil Biol. Biochem., 29 (4), 725-730.

Frederickson, J., Butt, K.R., Morris, R.M. and Denial, C. (1997). Combining vermiculture with traditional green waste composting systems. Soil Biol. Biochem., 29 (4), 725-730.

Fulekar and Jadia, 2008).

Fulekar, M.H. and Jadia, C.D. (2008). Phytoremediation: The application of vermicomposts to remove zinc, cadmium, copper, nickel and lead by sunflower plant. Environmental Engineering and Management Journal, 7(5), 547-558.

Gajalakshmi, S., Ramasamy, E.V. and Abbasi, S.A. (2001). Assessment of sustainable vermiconversion of water hyacinth at different reactor efficiencies employing *Eudrilus eugeniae* (Kinberg). Biores. Technol., 80, 131-135.

Gajalakshmi, S., Ramasamy, E.V. and Abbasi, S.A. (2002). Vermicomposting of paper waste with the anecic earthworm *Lampito mauritii* (Kinberg). Indian J. Chem. Technol., 9, 306-311.

Gajalakshmi, S., Ramasamy, E.V. and Abbasi, S.A. (2002a). Vermicomposting of paper waste with the anecic earthworm *Lampito mauritii* (Kingberg). Indian J. Chem. Technol., 9, 306-311.

Gajalakshmi, S., Ramasamy, E.V. and Abbasi, S.A. (2002b). Vermicomposting of different forms of water hyacinth by earthworm *Eudrilus eugeniae,* (Kinberg). Bioresour. Technol., 82, 165–169.

Gajalakshmi, S., Ramasamy, E.V. and Abbasi, S.A. (2005). Composting-Vermicomposting of leaf litter ensuring from the trees of mango (*Mangifera indica*). Bioresour. Technol., 96, 1057-1061.

Gamaley, A.V., Nadporozhskaya, M.A., Popov, A.I., Chertov, O.G., Kovsh, N.V. and Gramova, O.A. (2006). Non-root nutritional with vermicompost extract as the way of ecological optimization, plant nutrition-food security and susceptibility of argo-ecosystem. World Journal of Hort., pp. 862-863.

Garcia, M., Davared, C., Gallego,P and Toumi, M. (1999). Effect of various calcium-potassium ratio on cation nutrition of grapes grown hydroponically. J. Plnat Nutri., 22, 417-425.

Garg, K. and Bhardwaj, N. (2000). Effect of vermicompost of Parthenium on two cultivers of wheat. Indian J. Ecol., 27, 177-180.

Garg, P., Gupta, A., and Satya, S. (2006 a). Vermicomposting of different types of waste using *Eisenia foetida*: A comparative study. Biores. Technol., 97, 391–395.

Garg, V.K. and Kaushik, P. (2005). Vermistabilization of textile mill sludge spiked with poultry droppings by an epigeic earthworm *Eisenia foetida.* Bioresource Technology, 96, 1063-1071.

Garg, V.K. and Kaushik, P. (2005). Vermistabilization of textile mill sludge spiked with poultry droppings by an epigeic earthworm *Eisenia foetida*. Bioresource Technology, 96, 1063-1071.

Garg, V.K., Chand, S., Chhillar, A., and Yadav, Y.K. (2005). Growth and reproduction of *Eisenia foetida* in various animal wastes during vermicomposting. Applied Ecol. and Environ. Res., Hungary, 3 (2), 51-59.

Garg, V.K., Yadav, Y.K., Sheoran, A., Chand, S. and Kaushik, P. (2006 b). Live stocks excreta management through vermicomposting using an epigeic earthworm *Eisenia foetida*. Environmentalist, 26, 269-276.

George, S., Giraddi, R.S., and Patil, R.H. (2007). Utility of vermiwash for the management of thrips and might on chili (*Capsicum annum* L.) amended with soil organics. Karnataka Journal of Agricultural Science, 20, 657-659.

Ghatnekar, S.D., Mahavash, F.K., Ghatnekar, G.S. and Ghatnekar, M.S. (1998). Ecotechnology for pollution control and environmet management. Eviro Media, Pp. 59-67.

Ghosh, C. (2004). Integrated Vermi-Pisciculture: An alternative option for recycling of solid municipal waste in India. Bioresource Technology, 93, 71-75.

Ghosh, M., Chattopadhyay, G.N. and Baral, K. (1999). Transformation of phosphorus during vermicomposting. Biores. Technol., 69, 149-154.

Giraddi, R.S. (2001). Method of extraction of earthworm wash, a plant promoter substance, Souvenir and Abstracts. A paper presented in Silver Jubilee Celebration of Indian Society of soil of biology and ecology and VII$^{th}$ National Symposium on soil biology and ecology held at Bangalore during Nov. 7-9, 2009 pp. 66.

Gnanasoundari, A. (1992). Treatment of activated sludge using earthworm *Lampito mauritii*. M.Sc. Thesis, Annamalai University, Madras.

Godase, S.K. and Patel, C.B. (2001). Studies on the influence of organic manures and fertilizer doses on the intensity of sucking pest infesting brinjal. Pl. Pro. Bull., 53, 10-12.

Govindan, V.S. (1998). Vermiculture and vermicomposting in ecotechnology for pollution control and environment management. (R.K. Trivedy and A. Kumar eds.). Environ. Media. Karad, 49-57.

Graff, O. and Markeschin, F. (1980). Beeinflussuing des ertrags vonwiedelgras (*Lolium multiflorum*) durch Ausscheidungen von Regenwiirmern dreie verschiedener Arten. Pedobiologia, 20, 176-180.

Grappelli, A., Galli, E. and Tomati, U. (1987). Earthworm casting effect on *Agaricus bisporus* fructification. Agrochemical, 21, 407-416.

Gratelly, P., Benitez, E., Elvira, C., Polo, A. and Nogales, R. (1996). Stabilization of sludges from a dairy processing plant using vermicomposting. (In: C. Rodriquez-Barreueco, eds.) Fertilizers and Environment, Kluwer. The Netherlands, pp. 341-343.

Grundon, N.J. (1980). Effectiveness of soil dressing and foliar spray of copper sulphate in correcting copper deficiency of wheat (*Triticum aestivum*) in Queensland. Australian Journal of Experimental Agriculture and Animal Husbandry, 20, 717-723.

Gunadi, B. and Edwards, C.A. (2003). The effect of multiple applications of different organic wastes on the growth fecundity and survival of *Eisenia foetida* (Savigny) (Limbricidae). Pedobiologia, 47 (4), 321-330.

Gunadi, B., Blount, C. and Edwards, C.A. (2002). The growth and fecundity of *Eisenia foetida* (Savigny) in cattle Solid pre composted for different periods, Pedobiologia 46, 15-23.

Gupta, P.K. (2005). Vermicomposting for sustainable agriculture. Bharat Printing Press, Jodhpur, India, pp.11-14.

Gupta, R. and Garg, V.K. (2007). Stabilization of primary sludge during vermicomposting. J. Hazard Matter. doi:10.1016/j.hazmat.09.055.

Gupta, S.K., Tewari, A., Srivastava, R., Murthy, R.C. and Chandra, S. (2005). Potential of *Eisenia foetida* for sustainable and efficient vermicomposting of fly ash. Water Air Soil Pollut., 163, 293–302.

Haimi, J. and Hutha, V. (1988). Capacity of various organic residues to support adequate earthworm biomass forvermicomposting. Biolo. Fertil. Soil, 2, 23-27.

Halim, M., Conte, P., Piccolo, A. (2003). Poptential availability of heavy metals to phytoextraction from contaminated soils induced by exogenous humic acid substances. Chemoshere, 52, 265.

Hammermeister, A.M., Warman, P.R., Jeliazkova, E.A. and Martin, R.C. (2004). Nutrient supply and lettuce growth in response to vermicomposted and composted cow manure. Submitted to Bioresource Technology, Dec, 2004.

Hand, P. Hayes, W.A., Satchell, J.E. and Frankland, J.C. (1988). The vermicomposting of cow slurry. Pedobiologia, 21, 199-209.

Hartenstein, R., Neuhauser, E.F., and Kaplan, D.L. (1979). A progress reports on the potential use of earthworms in sludge management, In proceedings of the English National Sludge Conference, Florida. Information Transfer Inc. Silver Spring MD., pp. 238-241.

Herzenjak, T., Herzenjak, M., Kasuba, V., Marinculic, P.E. and Levanat, S. (1992). A new source of biologically active compounds in earthworms tissue (*Eisenia foetida, Lumbricus rubellus*). Comp. Biochem. Physiol., 102, pp. 441.

Herzenjak, T., Popovic, M. and Rudman, L.T. (1998). Fibrinolytic activity of earthworms extract (G-90) on lysis of fibrin clots originated from the venous blood of patients with malignant tumors. Pathol. Oncol. Res., 4(3), 201-241.

Hoffland, E., Jeger, M.J. and Van Beusichem, M.L. (2000). Effect of nitrogen supply on disease resistance in tomato depends on the pathogen, Plant and Soil, 218, 239-247.

Ismail, S. A. (1997). Vermicology: The biology of Earthworms, Orient Longman, press, Hyderabad, pp, 92.

Ismail, S.A. (1993). Keynote Papers and Extended Abstracts. Congress on traditional sciences and technologies of India, I.I.T., Mumbai, 10, 27-30.

Ismail, S.A. (1994). Use of local species of earthworms in vermicomposting and dissemination of the technology among farmers. Paper presented in National meeting on waste recycling, Centre of Science of Villages, Wardha, pp. 2-6.

Ismail, S.A. (1995). Earthworm in soil fertility, management and organic agriculture. (In: P.K. Thompson, eds.), Peekay Tree Crops development Foundation, Cochin, India, pp. 77-100.

Jadhav, A.D., Talasilkhar, S.C. and Power, A.G. (1997). Influence of the conjugative use of FYM, vermicompost and urea on growth and nutrient uptake in rice. Journal of Maharashtra Agriculture Universities, 22 (2), 249-250.

Jayashankar, S. (1994). Studies on vermicomposting as a method of sewage sludge disposal, M.Sc. Thesis, Anna. Univer. Madras.

Joshi, R. and Vig, A.P. (2010). Effect of vermicompost on growth, yield and quality of Tomato (*Lycopesicum esculenum* L.). African J. Basic and Appl. Sci., 2(3–4): 117–123.

Kale, R.D. (1991). Vermiculture scope for new biotechnology. (In: A.K.Ghosh eds.) Earthworm Resources and Vermiculture. Zoological Survey of India, Calcutta, 105-108.

Kale, R.D. (1998). Earthworms: Nature's gift for utilization of organic wastes. (In: C.A. Edwards, eds.), Earthworm Ecology. Soil and Water Conservation Society. Ankeny, Lowa St. Lucie Press, New York, pp. 355–373.

Kale, R.D. and Bano, K. (1988). Earthworm culturing techniques for production of Vee. Comp 83E. UAS Vee Meal 83P. UAS Mysore J. Agric Sci., 22 (3), 339-342.

Kale, R.D. and Krishnamoorthy, R.V. (1979) A comparative account of coelomocyte and haemocyte of five species of earthworms. Proc Indian Acad Sci., 88, 329-337.

Kale, R.D., and Bano, K. (1991). Time and space relative population growth of *Eudrilus eugeniae.* (In: G.K.Veeresh, D. Rajagopal, C.A. Virakamath, eds.): Advances in Management and Conservation of Soil Fauna, Oxford and IBH, New Delhi, pp. 657-664.

Kalra, N., Jain, M.C., Joshi, H.C. and Kumar, V. (1997). Fly ash as a soil conditioner and fertilizer. Biores. Technol., 64, 163-167.

Karmegam, N. and Daniel, T. (2000). Utilization of some weeds as substrates for vermicompost perparation using epigiec earthworm *Eudrilus euginiae.* Asian J. Microbial Biotech. Env. Sci., pp. 263-266.

Karmegam, N., Algumalai, K. and Daniel, T. (1999). Effect of vermicompost on the growth and yield of green gram (*Phaseolus aureus* Roxb.). Tropical Agriculture (Trinidad), 76, 143-146.

Karuna, K., Patil, C.R., Narayanswamy, P. and Kale, R.D. (1999). Stimulatory effect of earthworm body fluid (vermiwash) on crinkle red variety of *Anthurium andreanum* lind. Crop. Res., 17(2), 253-257.

Kaushik, P. and Garg, V.K. (2003). Vermicomposting of mixed solid textile mill sludge and cow dung with the epigeic earthworm *Eisenia foetida*. Bioresource Technol., 90, 311- 316.

Kaushik, Priya and Garg, V. K. (2003). Vermicomposting of mixed solid textile mill sludge and cow dung with the epigeic earthworm *Eisenia foetida*. Bioresource Technol., 90, 311- 316.

Kaushik, Priya and Garg, V. K. (2004). Dynamic of biological and chemical parameters during vermicomposting of solid textile sludge mixed with dung and agricultural residues. Bioresource Technol., 94 (2), 203-209.

Kaviraj and Sharma, S. (2003). Municipal solid waste management through vermicomposting employing exotic and local species of earthworms. Bioresource Technol., 90, 169-173.

Klock- Moore, K. A. (2001). Effect of controlled- released fertilizer application rates on bedding plant growth in substrate containing compost. Compost Science and Utilization, 9, 215-220.

Kobatke, M. (1954). The antibacterial substances extracted from lower animal. The earthworms Kekkaby (Tuberculosis), 29, 60-61.

Kpomblekou, A.K., Ankumah, R.O. and Ajwa, H.A. (2002). Trace and nontrace element content of broiler litter. Communications in Soil Science and Plant Analysis, 33, 1799-1811.

Krishnamoorthy, R.V. and Vajranabhiah, S.N. (1986). Biological activity of earthworm casts: An assessment of plant growth promoter levels in casts. Proceedings of the Indian Academy of Science (Animal Science), 95, 341-350.

Kumar, K. and Singh, K.P. (2001). Enriching vermicompost by nitrogen fixing and phosphate solubilizing bacteria. Biores. Technol., 76, 173-175.

Lasat, M.M. (2002). Phytoextraction of toxic metals: A review of biological mechnasims. J. Environ. Qual., 31, 109-120.

Ledesma, A., Arguello, J.A., Nunez, S.B. and Nieto, N. (2000). Effect of $GA_{3,}$ CCC and nitrogen fertilization on bulbing in garlic. In: Alliums 2000. 3rd Intl. Symp. Edible. Alliacaece. Athens, Ga 29 Oct.- 3 Nov., 2000.

Levi-Minzi, R., Riffalda, R. and Saviozzi, A. (1986). Organic matter and nutrients in fresh and mature farmyard manure. Agric.Wastes, 16, 225-236.

Loh, T.C., Lee, Y.C., Liang, J.B. and Tan, D. (2005). Vermicomposting of cattle and goat manures by *Eisenia foetida* and their growth and reproduction performence. Bioresource Technol., 96, 111-114.

Lowe, C.N. and Butt, K. R. (2002). Growth of hatchling earthworm in the presence of the adults: Interaction in laboratory culture- Biology and Fertility of Soil, 35, 204-209.

Lozek, O. and Fecenko, J. (1998). Effect of organo mineral fertilizer vermisol special on the quality and quantity of winter wheat yield. Folia-Universitatis Agriculturae Stetinesis, Agricultura, 72, 185-189.

Lozek, O. and Gravova. (1999). The influence of virmisol on the yield and quality of tomatoes. Acta. Horticulturae et-Regiotecturae, 2 (1), 17-19.

Maboeta, M.S. and Van-Rensburg, L. (2003). Vermicomposting of industrially produced woodchips and sewage sludge utilizing *Eisenia foetida*. Ecotoxicol. Environ. Safety, 56, 256–270.

Madjen, E., Burgos, P., Murilo, J. M. and Cabrera, F. (2001). Phyto-toxicity of organic amendments on activities of selected soil enzymes. Communications in Soil Science and Plant Analysis, 32, 2227-2239.

Maheshwarappa, H. P., Nanjappaand, H. V. and Hedg, M. (1999). Influence of organic manure on yield of arrowroot, soil physico-chemical and biological properties when grown as intercrop in coconut garden. Annals. of Agricul. Res., 20 (3), 318-323.

Mall, A.K., Dubey, A and Prasad, S. (2005). Vermicompost: An inevitable tool of organic farming for sustainable agriculture. Agrobios Newsletter, 3(8), 10-11.

Manivannan, S. (2004). Standardization and nutrient analysis of vermicomposting sugarcane wastes, pressmud-trash-bagasse by

*Lampito mauritii* (Kingberg) and *Perionyx exacavatus* (Perrier) and the effect of vermicomposts on soil fertility and crop productivity. Ph. D., Thesis, Annamalai University, India.

Manning, R. (2000). Food's Frontier: The next Green Revolution, North Point Press, New York, pp. 240.

Marinari, S., Masciandaro, G., Ceccanti and Grego, S. (2000). Influence of organic and mineral fertilizers on soil biological and physiological properties, Bioresource Technol., 72, 9-17.

Marschnar, H. (1995). Mineral nutrition of higher plants. Second Edition, Accademic press Londan., pp. 285-299.

Martin, J.P., Black, J.H. and Hawthorne, R.M. (1976). Earthworm biology and production, leaflet 2828. University of California Cooperative extension service.

Masciandaro, G., Ceccanti, B and Garcia, C. (2000). "In situ"vermicomposting of biological sludge's and impacts on soil quality. Soil Biology and Biochemistry, 32,1015-1024.

Mba, C.C. (1996). Treated-cassaca peel vermicomposts enhanced earthworm activities and Cowpea growth in field plots. Resources, Conservation and Recycling, 17, 219-226.

Mba, C.C. (1997). Treated-cassaca peel vermicomposts enhanced earthworm activities and cowpea growth in field plots. Resources, Conservation and Recycling, 17, 219-226.

Mitchell, A. (1997). Production of *Eisenia foetida* and vermicompost from feedlot cattle manure. Soil Biol. Biochem., 29, 763-766.

Mulongoy, K.and Bedoret, A. (1989). Properties of worm casts and surface soil under various plant covers in the humid tropics, Soil Biol. Biochem., 21, 197-203.

Muthukumaravel, K., Amsath, A and Sukumaran, M. (2008). Vermicomposting of vegetable wastes using cow dung. E. Journal of Chemistry, 5 (4), 810-813.

N' Dayegamiye, A. (1990). Effects of a long term d' apport defumier solide de bovins sur I' evolution des caracteristiques chimiques due sol et de la production de maisensilage, Canadian J. Plant Sci., 70, 767-775.

Nagasawa, H., Sawaki, Fujii, Y., Kobayashi, M., Segawa. T., Suzuki, R. and Inatomi, H. (1991). Inhibition by lumbricine from earthworm (*Lumbricus terrestris*) of the growth of spontaneous mammary tumors in SHN mice. Anticancer Res. II., pp. 1061.

Nagavallemma, K. P., Wani, Stephane, S. P., Lacroix, Padmaja, V. V., Vineela, C., Rao, M. B. and Sarawat, K. L. (2004). Vermicomposting; Recycling of wastes into valuable organic fertilizer. Global Theme on agroecosystem Report no. 8. Patancheru- 502324.

Nakasone, A.K., Bettiol, W., and De-Souza, R.M. (1999). The effect of water extracts of organic matter on plant pathogens, Summa Phytopathologica, 25, 330-335.

Nath, G. and Singh, K. (2011). Effect foliar spray of pesticides and vermiwash of animal, agro and kitchen wastes on soyabean (*Glycine max* L.). Crop Bot. Res. Int., 4(3): 52–57.

Nath, G. and Singh, K. (2012). Effect of vermiwash of different vermicompost on the Kharif crops. J. Central European Agri., 13(2): 379–402.

Ndegwa, P.M. and Thompson, S.A. (2001). Integrating composting and vermicomposting treatment and bioconversion of biosolids. Biores. Technol., 76, 107-112.

Ndegwa, P.M., Thompson, S.A. and Das, K.C. (2000). Effects of stocking density and feeding rate on vermicomposting of biosolids. Bioresour. Technol., 71, 5-12.

Nelson, D.W. and Sommers, L.E. (1982). Total organic carbon and organic matter. (In: A.L. Page, R.H. Miller and D.R. Keeney eds. Method of Soil Analysis), American Society of Agronomy, Medison, pp. 539-579.

Nelson, P.N. and Oades, J.M. (1998). Organic matter, sodicity and soil structure. (In: M.E. Sumner, and R. Naidu, eds. Sodic Soils), Oxford University Press, New York, pp, 76-91.

Nethra, N. N., Jayaprasad, K. V. and Kale, R. D. (1999). China aster (*Callistephus chinensis* (L) cultivation using vermicompost as organic amendent. Crop Research, Hisar, (2), 209-215.

Neuhauser, E.F., Loehr, R.C. and Makecki, M.R. (1988). The potential of earthworms for managing sewage sludge. (In: C.A. Edwards, E.F.

Neuhauser eds. Earthworms in wastes and environmental management), SPB Academic Publishing, The Hague, pp. 9-20.

Nogales, R., Elvira, C., Benitez, E., Thompson, R. and Gomez, M. (1999). Feasibility of vermicomposting dairy biosolids using a modified system to avoid earthworm mortality. J. Environ. Sci. Health, B34 (1), 151-169.

Orlikowski, L.B. (1999). Vermicompost extract in the control of some soil borne pathogens. International Symposium on Crop Protection, 64, 405-410.

Parr, J.F., Hornick, S.B. and Papendick, R.I. (2002). Transition from conventional agriculture to natural farming system: The role of microbial inoculants and biofertilizer. http.//www.emtech.org/data/pdf/0/03.pdf.

Parthasarathi, K. and Ranganathan, L. S. (2002). Supplementation of pressmud vermicast with NPK enhances growth and yield in leguminous crops (*Vigna munga* and *Arachis hypogaea*). J. Curr. Sci., 2, 35-41.

Parthasarathi, K. and Ranganathan, L.S. (2000). Aging effect on enzyme activities in pressmud vermicast of *Lampito mauritii* (Kinberg) and *Eudrilus eugeniae* (Kinberg). Biology and Fertility of Soil, 30, 347-350.

Pascual, J.A., Garcia, C., Hernndez, T. and Ayuso, M. (1997). Changes in the microbial activity of an arid soil amended with urban organic wastes. Biology and Fertility of Soil, 24, 429-434.

Pathak, R.K. and Ram, R.A. (2004). Manual on Jaivik Krishi, Central Institute for Subtropical Horticulture, Rehmankhera, P.O. Kokari, Lucknow-227107, 24, 31-32.

Payal, G., Gupta, A. and Satya, S. (2006). Vermicomposting of different types of wastes using *Eisenia foetida* a comparative study. Biores. Technol., 97, 391-395.

Peng, K., Li, X., Luo, C. and Shen, Z. (2006). Vegetation composition and heavy metals uptake by wild plants at three contaminated sites in Xiangxi area, China. Journal of Science and Health, Part A 40, 65-76.

Peyvast, G., Olfati, J.A., Madeni, S. Forghani, A., and Samizadeh, H. (2008). Vermicompost as a soil supplement to improve growth and yield of parsley. International Journal of Vegetable Science, 14(1), 82-92.

Popovic, M., Gradisa, M. and Harzenjak, T.M. (2005). Glycoprotein (G-90) obtained from the earthworms *Eisenia foetida* exerts antibacterial activity. Veterinarski Arhiv., 75(2), 119-128.

Popovic, M., Harzenjak, T.M., Babic, T., Kos, J. and Gradisa, M. (2001). Effect of earthworms (G-90) extract on formation and lysis of clot originated from venous blood of dogs with cardiopathies and malignant tumors. Pathol. Oncol. Res., 7, pp. 197.

Prakash, M. (2006). Antiulceral and anti-oxidative properties of earthworm paste of *Lampito mauritii* (Kinberg) on *Rattus norvegicus*. M. Phil., Thesis, Annamalai University.

Pramoth, A. (1995). Vermiwash: A potent bio-organic liquid "Ferticide". M.Sc. Thesis, University of Madras, India, pp. 29.

Purohit, S.S. (2003). Editorial Vermicomposting. A boon for soil health. Agrobios, News Letter, April 2003, pp. 3.

Rajkhowa, D.J., Gogoi, A.K., Kandal, R. and Rajkhowa, K.M. (2000). Effect of vermicompost on green gram nutrition. J. Indian Soc. Soil Sci., 48, 207-208.

Ramamoorthy, P. (2004). Standardization and nutrient analysis of vermicomposting sugarcane wastes, press-mud trash, bagasse by *Eudrilus eugeniae* (Kinberg) and *Eisenia foetida* (Savigny) and the effect of vermicompost on soil fertility and crop productivity. Ph.D. Thesis, Anna. Univ., India.

Ramamurthy, V. (2006).Vermicompost application improves the productivity and quality of Nagpur mandarain (*Citrus reticulata* Blanco). Organic News, (3), 1-3.

Rana., A.S. (2000). Effect of organic manures and foliar spray of nutrients and growth regulators on the growth and yield of soybean (*Glycine max.* L.) Merril. M.Sc. (Ag) Thesis submitted to Tamilnadu Agriculture University, Coimbatore.

Ranganathan, L.S. (2006). Vermibiotechnology from soil health to human health. Agrobios, India.

Rao, B.R.C. (2005). Vermicomposting. IEC CELL-KUDCEMP, Mysore.

Rao, N.S. (1994). Model scheme on financing vermicompost. National Bank Newslett., 5,1-5.

Raviv, M., Zaidman, B. and Kapulnik, Y. (1998). Compost Science and Utilization, 6, 46-52.

Reddy, B.G. and Reddy, S. (2000). Soil health and crop yields under organic farming in maize and soybean cropping system. In Fifth Symposium on soil and water management held under APAU, Rajendra Nagar, India, pp. 121-124.

Reganold, J. P., Glover, J. D., Andrews, P. K. and Hinman, H. R. (2001). Sustainability of three apple production systems. Nature J., 410, 925-926.

Reinecke, A.J., Viljoen, S.A. and Saayman, R.J. (1992). The suitability of *Eudrilus eugeniae, Perionyx excavatus and Eisenia foetida* (Oligochaeta) for vermicomposting in Southern Africa in terms of their temperature requirements. Soil Biol. Biochem., 24 (12), 1295-1307.

Reynolds, J.W. and Reynolds, W.M. (1972). Earthworms in medicine. Am. J. Nursing, 72, 1273.

Rodriguez, J.A., Zavaleta, E., Sanchez, P. and Gonzalez, H. (2000). The effect of vermicompost on plant nutrition, yield and incidence of root and crown rot of gerbera (*Gerbera jamesonii* H Bolus). Fitopathologia, 35, 66 -79.

Sagar, D. V., Naik, S. N. and Vasudevan. P. (2002). Vermicomposting of kitchen wastes and its effect on the quality and quantity of essential oil of *Ocimum sanctum* L. Proc. Intl. Composting and Compost Sci. Symposium, Columbus, Ohio: CD ROM.

Satchell, J. E. (1963). Nitrogen turnover by woodland population of *Lumbricus terrestris.* Soil Organisms, (J. Doeksen, and J. Van de Drift, eds) North Holland Publishing Co. Amsterdam, pp.60-66.

Scheuerell, S.J. and Mahaffee, W.F. (2002). Compost tea: Principal and of prospects for plant disease control. Compost Science Utilize., 10, 313-338.

Scott, M.A. (1988). The use of worm-digested animal wastes as a supplement to peat in loamless composts for hardy nursery stock. In Edwards, C.A., Neuhauser, E.F. (eds.), Earthworms in Environment and Wastes Managemnt. SPB Academic Publishing, The Netherlands, 211-220.

Senapati B.K. Kale, R.D. and Dash, M.C. (1984). Proceeding of the national seminar on organic waste utilization and vermicomposting part (B) December 5-8.

Senapati, B.K., Dash, M.C., Rane, A.K. and Panda, B.K. (1980). Observation on the effect of earthworms in the decomposition process in soil under laboratory conditions. Com. Physiol. Ecol., 5,140-142.

Setboonsarng, S. and Gilman, J. (1999). Alternative agriculture in Thailand and Japan, (http://www.solution-site.org/artman/publish/article_15shtml)

Sharma, D., Katnoria, J.K and Vig, A.P. (2011). Chemical changes of spinach waste during composting and vermicomposting. African J. Biotech., 10(16): 3124–3127.

Shield and Earl, B. (1982) Raising earthworms for profit. Shields Publication. P.O. Box 669 Eagle River Wisconsin, pp. 128.

Shivsubramanian, K. and Ganeshkumar, M. (2004).Influence of vermiwash on biological productivity of marigold. Madras Agriculture Journal., 91, 221-225.

Shukla, R.C. and Singh, K. (2013). Vermicomposts: A alternative biofertilizers for Zayad crops. J. Agri. Technol. 9(3): 711–726.

Shuster, W.D., Subler, S. and McCoy, E.L. (2000). Foraging by deep-burrowing earthworms degrades surface soil structure of a fluventic Hapludoll in Ohio. Soil Tillage Res., 54, 179-189

Shweta., Singh, Y.P. and Kumar, K. (2004) Vermicomposting – A profitable alternative for developing country. Agrobios Newsletter, 3 (4), 15-16.

Siminis, C.I., Loulakis, M., Kefakis, M., Manios, T. and Manios, V. (1998). Humic substances from compost affect nutrient accumulation and fruit yield in tomato. Acta Horticulturae, 469, 353-358.

Singh A, Singh DK (2001) Molluscicidal activity of Lawsonia inermis and its binary and tertiary combinations with other plant derived molluscicides. Ind J Exp Biol 39:263–268.

Singh VK, Singh DK (1995) Characterization of allicin as a mollus- cicidal agent in Allium sativum (Garlic). Biol Agr Hort 12:119–31.

Singh, D.K., Singh, A. (1993). Allium sativum (Garlic), a potent new Molluscicide. Biol. Agric. Hortic. 9, 121–124.

Singh, V.K. and Singh, D.K. (1996) Enzyme inhibition by allicin, the molluscicidal agent of Allium sativum L. (garlic). Phytotherapy Research, 10, 383–386.

Sligh, M. (2002). Organic at the Cross- roads: The past and future of organic movement. In A. Kimbrell (ed) Fatal Harvest: The Tragedy of Industrial Agriculture, Island Press, Washington, 341-345.

Snieg, L. and Bury, M. (1998). The result of application bio-preparation in the cultivation of potato and winter oil seed rape. Part-I. The influence of selected biopreparation on potato growth and development. *Folia-Universitatis, Agriculturae.* Stetinensis; Agricultura, 72, pp. 311-317.

Sokal, R.R. and Rohlf, F.J. (1973). Introduction of biostatistics. W. H. Freeman and Co. San Francisco.

Soytong, K. and Soytong, K. (1996). *Chaetomium* as a new broad spectrum mycofungicidde. Proc. of the First International Symposium on Biopesticides, Thailand, October 27-31, pp,124-132.

Soytong, K., Usuwan, P., Kamokmedhakul, S., Kanokmedhakul, K., Kukongviriyapan, V. and Isobe, M. (1999). Integrated biological control of *Phytophthora* rot of sweet orange using mycofungicides in Thailand. Proc. of the 5th International Conference on Plant Protection in the Tropics, 15-18 March, Malaysia, pp.329-331.

Sreenivas, C., Murlidhar, S. and Rao, M. S. (2000). Vermicompost, a viable component of IPNSS in nitrogen nutrition of ridge gourd. Annals of Agricultural Research, 21(1), 108-113.

Srivastava, A. K. and S. Singh. (2004). Nutrient diagnostics and management in citrus. Tech. Bulletin-8, Published by NRC for Citrus, Nagpur. Pp-130.

Stephenson, J. (1930). In oligochaeta claredon Press Oxford, pp. 658.

Stevinus, G., and Dunn, D. (2004). Fly ash a lining material for cotton. Journal of Environmental Quality, 33, 343-348.

Stone, A.G. (2002). Organic matter mediated suppression of *Pythium, Phytophthera,* and *Aphonomyces* root rots in field siols. Columbus, Ohio: CD Rom. Proc. Intl. composting and compost Sci. symposium,

Subasashri, M. (2004). Vermiwash: A effactive biopesticide. The Hindu News paper 30th sept. In Science and Technology section.

Subler, S. (1995). Earthworms in agro ecosystems. (In: P.F. Hendrix, ed. Earthworm Ecology and Biogeography in North America), Lewis Publisher, Boca Raton, FL,

Sudha, R., Ganesh, P., Mohan, M., Saleem, S. S., and Vijaylaxmi, G.S. (2003). Effect of vermiwash on the growth of black gram (*Vigna mungo)*. Agrobios News letter, 30(1), 77-79.

Suthar, S. (2007 c). Production of vermifertilizer from guar gum industrial waste by using composting earthworm *Perionyx sansibaricus* (Perrier). Environmentalist, 27 (1), 329–335.

Suthar, S. (2006 a). Potential utilization of Guar gum industrial wastes in vermicomposts productions. Biores. Technol., 97, 2474-2477.

Suthar, S. (2006 b). Vermiculture: taknique avum upyogita–Scientific Publishers Jodhpur, India.

Suthar, S. (2007 b). Nutrient changes and biodynamics of epigeic earthworm *Perionyx excavatus* (Perrier) during recycling of some agriculture wastes. Bioresour Technol., 98, 1608–1614.

Suthar, S. (2007 d). Influence of different food sources on growth and reproduction performance of composting epigeics: *Eudrilus eugeniae, Perionyx excavatus* and *Perionyx sansibaricus*. Applied Ecol. and Environ. Research, 5 (2), 79-92.

Suthar, S. (2007a). Microbial decomposition efficiencies of monoculture and polyculture vermireactors based on epigeic and anecic earthworms. World J. Microbiol Biotechnol., doi, 10.1007/511274-007-9635-9.

Suthar, S. (2008 a). Bioconversion of post harvest crop residues and cattle shed manure into value added products using earthworm *Eudrilus eugeniae,* (Kinberg). Ecological Engineering, 32, 206-214.

Suthar, S. (2008 b). Feasibility of vermicomposting in bio-stabilization of sludge from a distillery industry. Science of total Environment, 394, 237-243.

Suwan, S., Isobe, M., Kanokmedhakul, S., Lourit, N., Kannokmedhakul, K., Soytong, K. and Koga, N. (2006). Elucidation of great micro-hetrogirnity of an acidic neutral *Trichotoxin* mixture from *Trichoderma harziam* by ESI- QTOF. Mass Spectormetry, 35, 1438-1453.

Szczech, M., Rondomanski, W., Brzeski, M.W., Smolinska, U. and Kotowski, J.F. (2001). Suppressive effect of a commercial earthworm compost on some root infecting pathogens of cabbage and tomato. Biol. Agri. and Horti., 10 (1),47-52.

Szczech, M., Rondomanski, W., Brzeski, M.W., Smolinska, U. and Kotowski, J.F. (1993). Suppressive effect of a commercial earthworm compost on some root infecting pathogens of cabbage and tomato. Biol. Agri. and Horti., 10(1),47-52.

Talarposhti, R.M. and Kambouzia, J. (2007). Influence of organic and chemical fertilizers on growth and yield of tomato (*Lycopersicon esculantum* L.) and soil chemical properties, Tropentag, October 9-11-Witzenhausen.

Talashilkar, S.C. and Dosani, A.A.K. (2008). Earthworms in agriculture. Agrobios, India.

Talashilkar, S.C., Bhangarath, P.P. and Mehta, V.P. (1999). Changes in chemical properties during vermicomposting of organic residues as influenced by earthworm activity. J. Indian Soc. Soil Sci., 47, 50-53.

Talashilkar, S.C., Bhangarath, P.P. and Mehta, V.P. (1999). Changes in chemical properties during vermicomposting of organic residues as influenced by earthworm activity, J. Indian Soc. Soil Sci., 47, 50-53.

Taylor, M., Clarke, W.P. and Greenfield, P.F. (2003). The treatment of domestic waste water using small-scale vermicompost filter beds. Ecol. Eng., 21(2–3), 197–203.

Thangavel, P., Balagurunathan, R., Divakaran, J. and Prabhakaran, J. (2003). Effect of vermiwash and vermicast extraction soil nutrient status, growth and yield of paddy. Advances of plant Sciences. 16, 187-190.

Thompsons, R. B. and Nogales, R. (1999). Nitrogen and Carbon mineralization in soil of vermicomposted and unprocessed olive cake (*'Orujo seco*) produced from two stage centrifugation for olive oil extraction. J. Env. Sci. and Health, Part - B, Pesticides, Food Contaminants and Agricultural Wastes, 34 (5), 917-928.

Thomsen, I.K. (2001). Recovery of nitrogrn from composted and anaerobically stored manure labeled with $^{15}N$. European Journal of Agronomy,15, 31-41.

Tiwari, S.C., Tiwari, B.K. and Mishra, R.R. (1989). Microbial populations, enzyme activities and nitrogen-phosphorus-potassium enrichments in earthworm casts and in the surrounding soil of a pineapple plantation. Biol. Fertil. Soils, 8, 178-182.

Todkari, A.A. (2001). Effect of vermiwash prepared by two methods on the growth characteristic, yield and nutrition of three flowering plants. M.Sc. (Agri) Thesis Submilted to Dr. B.S.K.K.V. Dapoli.

Todkari, A.A. and Talashilkar, S.C. (2001). Effect of vermiwash prepared by two methods on the growth characteristics, yield and nutrition of three flowering plant. Souvenir and Abstracts. A paper presented in Silver Jubilee Celebration of the Indian Society of Soil Biology and Ecology and VII$^{th}$ National Symposium on Soil Biology and Ecology held at U.A.S. Bangalore. During 7-9 Nov. pp. 97.

Tomati, U. and Galli, E. (1995). Earthworms soil fertility and plant productivity, Acta Zoologica Fennica., 196, 11-14.

Tomati, U., Galli, E., Pasetti, L. and Volterra, E. (1995). Bioremediation of olive mill wastes waters by composting. Waste Manag. Res., 13, 509-518.

Tomati, U., Grappelli, A. and Galli, E. (1987). The presence of growth regulators in earthworm-worked wastes. (In: A.M.Bonvicini Paglioi and P. Omodeo, eds.). Proceeding of International Symposium on Earthworms. Selected Symposia and Monographs, Unione Zoologica Italiana, 2, Mucchi, Modena, pp. 423-435.

Tripathi, G. and Bharadwaj, P. (2004). Comparative studies on biomass production, life cycles and composting efficiency of *Eisenia foetida* (Savigny) and *Lampito mauritii* (Kinberg). Biores. Technol., 95, 77-83.

Trivedi, R.B. and Kumar, A. (1998). Ecotechnology for Pollution Control and Environmental Management. Eviro Media, PP. 196-197.

Trivedi, R.C. and Goel, D. (1984). Water pollution and Physiochemical properties II Edition, Pragati Prakashan Meerut, pp. 304-346.

Trobisch, S. and Schilling, G. (1970). Beitrag zur klarung der physiologischen grundlage der samenbil dung bei, einjahri-gen pflanzen und, zur, wirkung spatter zusatzlicher n- gbn auf deisen prozess am beis piel boon *Sinapis alba* L. Albrecht-Thaer Archive, 14, 253-265.

Tuitert, G., Szczech, M. and Bollen, G.J. (1998). Supression of *Rhizoctonia solani* in potting mixture amended with compost made from organic household wastes. Phytopathology, 88, 764-773.

Turner, M.S., Clark, C.A., Stanely, C.D. and Smajstrala, A.G. (1994). Physical characteristics of sandy soil amended with municipal solid waste compost, Soil Crop Sci. Florida Proc., 53, 24-26.

Umamaheswari, S., Viveka, S. and Vijayalakshmi, G.S. (2003). Indigenous vermiwash collecting device. The Hindu, Jul.17, 1-2.

Vadiraj, B.A., Siddagangaiah, D. and Potty, S.N. (1998). Response of *Coriandrum sativum* (L.) cultivars to graded levels of vermicompost. Journal of Spices and Aromatic Crops, 7(2), 141-143.

Vasanthi, D. and Kumaraswamy, K. (1999). Efficacy of vermicompost to improve soil fertility and rice yeilds. J. Indian Soc. Soil Sci., 47, 268-272.

Venkatesh, R.M. and Eevera, T. (2008). Mass reduction and recovery of nutrients through vermicomposting of fly ash. Appl. Ecol. and Environ. Res., 6(1), 77-84.

Verma, T.S. and Sharma, P.K. (2000). Effects of organic residue management on productivity of the rice-wheat cropping system. Long-term Soil Fertility Experiments in Rice-Wheat Cropping System (I.P., Abrol, K.F., Bronson, J.M. Duxbury, and R.K., Gupta, eds.), Rice-Wheat Consortium Series-6, pp. 163-171.

Vermi Co. (2001). Vermicomposting technology for waste management and agriculture; an executive summary. (http://www.vermico. com/ summary.htm), Po. Box 2334, Grants pass, OR, 97528, USA: Vermi Co.

Viallier, J., Cadoret, M.A., Roach, P. and Valembois, P. (1985). Protein analysis of earthworm coelomic fluids Isolation and characterization of several bacteriostatic molecules from *Eisenia foetida* and *Eisenia anderi*. Develop. Comp. Immunol., 9, 11-20.

Vinceslas-Akpa, M. and Loquest, M. (1997). Organic matter transformation in lignocellulosic waste products composted or vermicomposted (*Eisenia foetida andrei*): chemical analysis and 13C CPMAS NMR spectroscopy. – Soil Biology and Biochemistry, 29, 751-758.

Vig, A.P., Singh, J. Wani, S.H. and Dhaliwal, S.S. (2011). Vermicomposting of tannery sludge mixed with cattle dung into valuable manure using earthworm *Eisenia fetida* (Sarigny). Biore. Tech., 102(17): 7941–7945.

Viveka, S., Vijayalakshmi, G.S. and Palaniappan, R. (2005). Vermicomposted weeds on the growth and yield of Bhendi plant. Environment and Ecology, 23(1), 29-32.

Wani, S.P. (2002). Improving the livelihood: New partnership for win-win solutions for natural resource management. Paper submitted in the 2nd International agronomy Congress held at New Delhi, India during 26-30 November.

Wani, S.P. Rupela, O.P.and Leo, K.K. (1995). Sustainable agriculture in the semi-arid tropics through biological nitrogen fixation in grains legumes, plant and soil, 174, 29-49.

Weerasinghe, K.W.L.K., Mohotti, K.M., Herath, C.N., Sanarajeewa, A., Liyangunawardena, V. and Hitinayake, H.M.G.S.B. (2006) Biological and chemical properties of "Vermiwash" a natural plant growth supliment for tea, coconut and horticulture crops, 12 September Forestry and Environment Symposium, University of Jayewardenepura, Sri Lanka.

Weisbach, W.W. (1962). Regenwurmer and essbare erde. Biol. Jaarb. Dodonea, 30, 225-238.

Welki, S.E. and Parkinson, D. (2003). Effect of *Aporrectodea trapezoides* activity on seedling growth of *Pseudotsuga menziesii*, nutrient dynamics and microbial activity in different forest soils. Forest Ecol. Manage., 173, 169-186.

Weltzien, H.C. (1989). Some effects of compost organic materials on plant health. Agriculture Ecosystem and Environment, 27, 439-446.

Wezel, A. and Rath, T. (2002). Resource conservation strategy in agro-ecosystem of semi-arid West-Africa. Journal of Arid Environments, 51, 383-400.

Wong, S.H. and Griffiths, D.A. (1991). Vermicomposting in the management of pig-waste in Hong Kong. World. J. of Microbial. and Biotech., 7, 593-595.

Yadav, A.K. (2003). Organic forming- Trade Prectices and Regulation. RBDC Imphal.

Yadav, A.K., Kumar. K., Singh, S., Shweta, and Sharma, M. (2005). Vermiwash: A liquid biofertilizer, Uttar Pradesh J. of Zoology, 25(1), 97-99.

Yang, X.E., Peng, H.Y. and Jiang, L.Y. (2005). Phytoremediation of Copper from contaminated soil by *Elsholtzia splendeds* as affected by EDTA, citric acid and compost. International Journal of Phytoremediation, 7, 69-83.

Yegnanarayan, R., Ismail, S.A., and Shorti, D.S. (1998). Anti-inflammatory activity of two earthworm portions in carrageenan pedal oedema test in rats. Ind. J. Physiol. Pharmacol., 32, 72-74.

Yegnanarayan, R., Sethi, P.P., Rajhans, P.K., Pulandiran, K. and Ismail, S.A. (1987). Anti-inflammatory activity of total earthworm extract in rats. Ind. Pharmac., 19, 221- 224.

Yermiyahu, U., Keren, R. and Chen, Y. (2001). Effect of composted organic matter on boron uptake by plant. Soil Science Society of America Journal, 65, 1436-1441.

Zaller, J.G. (2006). Foliar spraying of vermicompost extracts. Effects on Fruit quality and indications of Late-Blight suppression of field grown tomatoes. Biol. Agri. and Horti., 24,165-180.

Zambare, V.P., Padul, M.V., Yadav, A.A. and Shete, T.B. (2008). Vermiwash: Biochemical and microbial approach as ecofriendly soil conditioner. ARPN Journal of Agriculture and Biological Science, 3(4), 1-5.

Zebrath, B.J., Neilsen, G.H., Hogue, E. and Neilsen, D. (1999). Influence des amendments faits de dechets organiques surcertains properties physiques et chemiques due sol. Canadian J. Soil Sci., 79, 501-504.

# Index

## K



## L



## M



## N



## O



## P

## R



## S



## T



## U



## V



## W